SpringerBriefs in Applied Sciences and Technology

Thermal Engineering and Applied Science

Series Editor

Francis A. Kulacki, University of Minnesota, USA

For further volumes:
http://www.springer.com/series/10305

Michael Ohadi · Kyosung Choo
Serguei Dessiatoun · Edvin Cetegen

Next Generation Microchannel Heat Exchangers

 Springer

Michael Ohadi
Department of Mechanical Engineering
University of Maryland
College Park, MD
USA

Kyosung Choo
Department of Mechanical Engineering
University of Maryland
College Park, MD
USA

Serguei Dessiatoun
Department of Mechanical Engineering
University of Maryland
College Park, MD
USA

Edvin Cetegen
Intel Corporation
Chandler, AZ
USA

ISSN 2193-2530   ISSN 2193-2549   (electronic)
ISBN 978-1-4614-0778-2   ISBN 978-1-4614-0779-9   (eBook)
DOI 10.1007/978-1-4614-0779-9
Springer New York Heidelberg Dordrecht London

Library of Congress Control Number: 2012947068

Preface

The goal of enhancing heat transfer while minimizing pressure drops and reducing the size and volume of energy conversion/thermal management systems has been the subject of intensive research for more than four decades. But growing energy demands, the need for increased energy efficiency and materials savings, space limitations for device packaging, and increased functionality and ease of unit handling have created revolutionary challenges for the development of high-performance, next-generation heat and mass exchangers. Among various techniques, innovative microchannel heat and mass exchangers appear to be the most promising way to meet these challenges in thermal management. When properly designed and utilized, microchannels can distribute the flow precisely among the channels, reduce flow travel length, and establish laminar flow in the channels while achieving high heat transfer coefficients, high surface area-to-volume ratios, and reduced overall pressure drops. These are among the major advantages of microchannels for use in a diverse range of industries. Current research on microchannels may represent only the tip of the iceberg of future possibilities for their expanded use. Rapid advancements in micro-machining and micro-deformation techniques are reducing the cost of fabrication while improving the reliability of microchannel systems, thus minimizing one of the main limitations of microchannels.

This book focuses on the latest developments in microchannel heat and mass exchangers, with a particular focus on presenting fundamental research results of practical significance. This book is divided into three chapters. Chapter 1 focuses on the fundamentals of microchannels, their classifications, fabrication techniques, and the design correlations of single-phase and two-phase heat transfer and pressure drops in microchannels. Chapter 2 introduces next-generation force-fed microchannels for high flux cooling applications, including the respective heat transfer and pressure drop characteristics of single-phase and two-phase flow systems. Finally, Chap. 3 discusses emerging applications of microchannels for both heat and mass transfer applications. Because one of the main objectives of this book is to introduce the latest information in the field, a conscious effort has been made to minimize coverage of related information that otherwise can be found in standard texts or technical references available in the open literature.

College Park, MD, USA   Michael Ohadi
College Park, MD, USA   Kyosung Choo
College Park, MD, USA   Serguei Dessiatoun
Chandler, AZ, USA   Edvin Cetegen

Acknowledgments

The authors' research on next generation microchannels and micro heat exchangers over the years has been supported by both the government and an industrial Consortium on Advanced Heat Exchangers and Electronics Cooling at the Center for Environmental Energy Engineering, as well as the CALCE Electronics Products and Systems Center at the University of Maryland, College Park. The authors would like to in particular acknowledge the support of Office of Naval Research (Contract No: N000140510539), with Dr. Mark Spector as the program manager, for enabling us to introduce new advancements to the field of force-fed microchannel heat exchangers in the past decade. We also would like to take this opportunity to acknowledge the support of Wolverine Tube, Inc. (Decatur, AL) and their MicroCool division for fabrication of some of the micro-grooved surfaces for testing in our micro heat and mass exchanger systems. In addition, this work was supported by the National Research Foundation of Korea Grant funded by the Korean Government (Ministry of Education, Science and Technology). [NRF-2012-357-2012R1A6A3A03038573]. Dr. Diradul Islam from the Petroleum Institute (Abu Dhabi) contributed to Chap. 1 of this manuscript (fundamentals of microchannels) and we hereby acknowledge his contributions.

Contents

Symbols

A	Heat transfer area (m^2)
Bo	Bond number
Bl	Boiling number
C_p	Specific heat (J/kg·K)
Co	Convection number
d_h	Hydraulic diameter (m)
F	Correlation factor
f	Friction loss coefficient
F_{fl}	Fluid-surface parameter
Fr	Froude number
G	Mass flux (kg/m^2·s)
g	Gravitational acceleration (m/s^2)
h	Heat transfer coefficient (W/m^{2}K) Enthalpy (J/kg)
$\bar{h}$	Average heat transfer coefficient (W/m^{2}K)
H	Height (mm)
h_{fg}	Latent heat (J/kg)
j	Total mixture volumetric flux (m/s)
k	Thermal conductivity (W/m·K)
L	Length (mm)
$\dot{m}$	Mass flow rate (kg/s)
n	Number of channels/tubes
Nu	Nusselt number
P	Pressure (N/m^2) Power (W)
Pr	Prandtl number
P_w	Perimeter (m)
q	Heat transfer rate (W)
Re	Reynolds number
S	Chen's suppression factor
t	Thickness (m)
T	Temperature (K)
$\bar{T}$	Average temperature (K)

V Fluid velocity (m/s)
w Width (m)
X Martinelli's factor
x Quality, length
z Distance (m)

Greek Symbols

Δ Difference
δ Film thickness
v Kinematic viscosity m^2/s
μ Fluid viscosity (kg/m·s)
ρ Fluid density (kg/m^3)
σ Surface tension (kg/s^2)
ϕ Two-phase frictional multiplier

Subscripts

app   Apparent
CBD   Convective boiling dominant
ch   Channel
FC   Forced convection component
f Saturated liquid
fd   Fully developed
fin   Fin
f0   All vapor-liquid mixture assumed to be saturated liquid
G Gas phase
g Saturated vapor
h Hydraulic diameter hydrodynamically; heated
i Inlet
L Liquid phase
L0   All mixture assumed to be liquid
lam   Laminar
NB   Nucleate boiling component
NBD   Nucleate boiling dominant
o Outlet
pump   Pumping
s Heated surface
t Thermally
TP   Two-phase

Chapter 1
Fundamentals of Microchannels

Abstract The growing demand for product miniaturization in all industrial sectors, coupled with global competition for more reliable, faster, and cost-effective products, has led to many new challenges for design and operation of thermal management systems. The rapid increase in the number of transistors on microchips, with increased functionality/power and consequently higher heat fluxes, is one such great challenge in the electronics packaging industry. Microchannel heat and mass exchanger technologies are finding new applications in diverse industries and emerging as a promising solution to game changing technologies in the way we design and operate next-generation, high performance thermal management systems. The discussion in this chapter will deal with fundamentals of microchannels. We begin by introducing the history, technical background, classification, advantages, and disadvantages of microchannels. Fabrication methods (conventional technology and modern technology) for microchannels are considered next. Finally, correlations of pressure drop and heat transfer coefficient for single-phase and phase-change flows are presented for a variety of internal flow conditions.

Keywords Fabrication · Micro heat and mass exchangers · Pressure drop · Heat transfer · Phase change · Two-phase · Correlations

1.1 Introduction and History of Microchannels

1.1.1 History

A great deal of work has been conducted on single-phase heat transfer in microchannels since Tuckerman and Pease's pioneering effort (1981) on the cooling of very large-scale integrated circuits (VLSI). In early 1981, Tuckerman and Pease

M. Ohadi et al., *Next Generation Microchannel Heat Exchangers*,
SpringerBriefs in Thermal Engineering and Applied Science,
DOI: 10.1007/978-1-4614-0779-9_1, © The Author(s) 2013

Fig. 1.1 Diagram showing scale of components and cooling system (Zang et al. 2003)

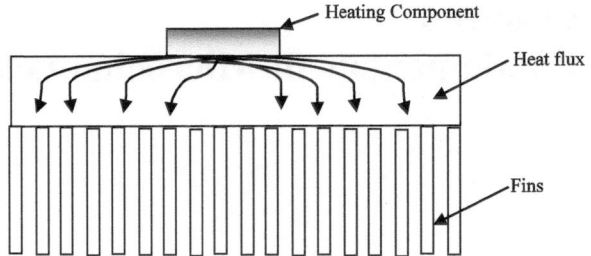

(1981) first explained the concept of microchannel heat sinks and predicted that single-phase forced convective cooling in microchannels could potentially remove heat at a rate of the order of 1,000 W/m^2.

Forced convection in channels and liquid injection has been used for faster and larger scale cooling in industry for decades. Microchannel heat transfer, however, has become increasingly popular and interesting to researchers due to high heat transfer coefficients, with potential for record-high heat transfer coefficient and low to moderate pressure drops when compared to conventional air and liquid cooled systems (Philips 1988; Gillot et al. 2000; Hsu et al. 1995; Hahn et al. 1997; Martin et al. 1995; Viday et al. 1993). For example, microchannel heat sinks have been demonstrated for high-power laser diode array cooling and have achieved a heat flux removal rate of 500 W/cm^2 (Missaggia et al. 1989; Mundinger et al. 1988; Beach et al. 1992). Figure 1.1 shows the proposed design of a two-phase microchannel heat sink fabricated on the backside of a flip chip-bonded IC chip (Zang et al. 2003). In the last few decades, studies on two-phase flow and heat transfer characteristics in microchannel flow passages have become increasingly important due to the rapid development of micro-devices used for various engineering applications, such as medical devices, high heat flux compact heat exchangers, and cooling of high-power density micro-electronics, super-computers, plasma facing components, and high-powered lasers. The continuing push toward more densely packed microchips may require greater heat dissipation than that typically provided by simple forced air-cooling systems. Liquid cooling using microchannels integrated with microchips is the next most attractive alternative (Schmidt 2003).

1.1.2 Introduction of Microchannels

In most cases when cooling requirements are over 100 W/cm^2 they cannot be easily met either by simple air-cooling or water-cooling systems. In many applications, to dissipate the high heat flux of the components, the required heat sinks must be larger than the components themselves, as shown in Fig. 1.1. Nevertheless, hot spots usually appear, and non-uniform heat flux levels are observed at the heat sink level. This has motivated researchers to develop new heat sinks that can

Fig. 1.2 Silicon microchannel heat sink with heating component for cooling (Gillot et al. 2001)

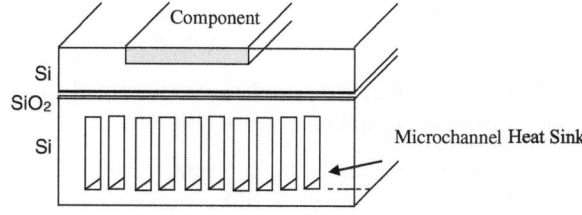

be directly embedded on the back of the heat source for uniform heat flux removal, as shown in Fig. 1.2. This heat sink is usually made out of silicon, with a silicon oxide layer to keep the component electrically insulated. Very narrow rectangular channels are formed with fins in the micrometer range that ensure uniform heat flux removal by circulating cold fluid through the rectangular microchannels.

From Newton's law of cooling we know that for a fixed temperature difference, heat flux depends on the product of hA, where h is the heat transfer coefficient and A is the heat transfer surface area. So, in order to fulfill the requirement of high heat flux removal the product of hA has to be increased, and since the heat transfer coefficient h is related to the hydraulic diameter, increasing surface area is one option. The heat transfer area can be increased by using microchannels on the heat generating body (chip surface), as shown in Fig. 1.2. The flow behavior of water inside the channels is determined by the channel hydraulic diameter and the channel cross-sectional area. To obtain high heat transfer, a smaller hydraulic diameter and a larger heat transfer area of the channel are preferred, so a number of narrow channels with high depth are suitable. Smaller hydraulic diameter and larger cross-sectional area increase the pressure drop and consequently require higher pumping power. On the other hand, increased cross-sectional area of the heating surface enhances the heat transfer rate. These requirements can be adjusted with next-generation microchannels that will have larger hydraulic diameter but provide larger cross-sectional area as well as high heat transfer coefficient.

1.1.3 Classification of Microchannels/Minichannels

We can classify microchannels in different ways. Several investigators have proposed different criteria for minichannels vs. microchannels. Serizawa et al. (2002) described one criterion for classification of microchannels proposed by Suo and Griffith (1964) as follows:

$\lambda \geq d_h$ where λ is the Laplace constant and d_h is the channel diameter.

Mehendale et al. (2000) used the hydraulic diameter to classify micro heat exchangers as follows:

- Micro heat exchanger: $1 \ \mu m \leq d_h \leq 100 \ \mu m$
- Meso-heat exchanger: $100 \ \mu m \leq d_h \leq 1 \ mm$

- Compact heat exchanger: $1\ \text{mm} \leq d_h \leq 6\ \text{mm}$
- Conventional heat exchanger: $d_h > 6\ \text{mm}$

Kandlikar (2002) proposed a microchannel classification for single-phase as well as two-phase applications as follows:

- Conventional channels: $d_h > 3\ \text{mm}$
- Minichannels: $3\ \text{mm}\ d_h > 200\ \mu\text{m}$
- Microchannels: $200\ \mu\text{m}\ d_h > 10\ \mu\text{m}$

Palm (2000) gave a more general definition of microchannels, describing them as heat transfer elements where the classical theories cannot correctly predict the friction factor and heat transfer. Stefan (2002) defined a microscale system as one whose typical phenomena are absent in a macro system. So it is not always suitable to differentiate mini- and microchannels by a specific diameter such as hydraulic diameter of 1 mm, although this definition is often used nonetheless.

1.1.4 Benefits and Challenges of Microchannels

Flow in microchannels has been widely investigated during the last two decades in the search for effective, faster cooling of high-power density electronic devices. As discussed above, the advantage of microchannels lies in their high heat transfer coefficient and ability to decrease the size of heat exchangers significantly. Other advantages are their reduced weight, low inventory, and reduced use of materials. The decreased diameters of microchannels result in more compact heat exchangers and higher heat transfer coefficients through more surface area per unit volume. Microchannels have wide practical applications in highly specialized fields, such as bioengineering and microfabricated fluidic systems, micropumps, and micro-heat pipes. For example, the compactness and low weight of microchannels have turned the automotive industry toward micro heat exchangers, and today micro-channels have almost completely replaced circular tubes in automotive condensers and heat exchangers with hydraulic diameters of around 1 mm. More recently, microchannels have been successfully applied to automotive air conditioning systems, fuel cells, and microelectronics. The main challenges of microchannels are the fabrication difficulties and high-grade filtering of the working fluid necessary for it to flow through the channels. High pressure drop and the pumping power required are also considered challenges of microchannels.

1.2 Fabrication Methods for Microchannels

Microchannels are fabricated by a variety of processes depending on the dimensions and materials used. Common materials used for microchannels are silicon, silica, polycarbonate/polyimide, plastic, or metal. Basic microchannel configurations are

Fig. 1.3 Taxonomic chart of
fabrication methods for
microchannels

Conventional Technology
{
Micro deformation
Micro sawing
Micro milling
Dicing
}

Modern Technology
{
MEMS
  - Wet etching
  - Deep RI-Etching
  - LIGA
  - Wafer bonding
Laser micro machining
Electro-discharge machining
Micromolding
}

rectangular, semicircular, triangular, or trapezoidal cross-sections, which are widely reported in the literature and are summarized by Nguyen and Werely (2002). Other geometrically complex microchannels may offer more attractive performance, but they have not yet been investigated. Since the first demonstration of microchannels by Tuckerman and Pease (1981), a number of microchannel fabrication methods have become standard processing approaches in this field. These methods can be divided into two groups, conventional technologies and modern technologies, as shown in Fig. 1.3. Conventional fabrication technologies include methods such as micro-deformation, micro-sawing, micro-milling, and dicing. Modern microchannel fabrication techniques include MEMS (Micro-Electro-Mechanical Systems) methods, laser micro-machining, electro-discharge machining, and micro-molding. MEMS technology has grown dramatically alongside semiconductor technology and is the most widely used technology in research laboratories. Recently, laser micro-machining technology has gained the spotlight due to the method's low manufacturing uncertainty and its potential to manufacture an unlimited number of geometries. This section will focus on micro-deformation, micro-sawing, deep RI-etching, and laser micro-machining technologies, which are widely used in manufacturing. Table 1.1 presents a summary of some fabrication methods for microchannels. Detailed explanations of the fabrication methods will be provided in the following sections.

1.2.1 Conventional Technology

1.2.1.1 Micro-Deformation

As indicated in Table 1.1, the micro-deformation technique can fabricate rectangular channels on any material. As reported by Kukowski (2003), the micro-deformation process can form up to 500 channels per inch. At this time, microchannels of up to 250 channels per inch are routinely formed on a broad

Table 1.1 Summary of some fabrication methods for microchannels

	Micro-deformation technology	Micro-machining	MEMS (Deep reactive ion etching)	Laser micro-machining
Geometries	Rectangular	Rectangular	Rectangular, circular, triangular, trapezoidal	Unlimited
Materials	Metal and non-metal	Metal and silicon	Metal, silicon, and glass	Metal and glass
Channel range	250 channels/ inch	0.1–10 mm	Nanometer scale to millimeter scale	Nanometer scale to millimeter scale
Advantages	Low cost, fast	High or low aspect ratio, inexpensive, fast	Low manufacturing uncertainty	Low manufacturing uncertainty
Disadvantages	Some materials require post-treatment	Complex design is impossible	Slow process (1 day)	Too expensive

range of materials. The channels are cut in one continuous pass or multiple passes depending on the system used. The advantages of the micro-deformation technology include low cost and quickness. However, depending on the strain-hardening rate of the materials, some processed materials after micro-deformation processing may require additional post treatment.

As shown in Fig. 1.4, the working principles of the micro-deformation technology are simple. Using the patented tool and prescribed interference angles with the work piece, the process plastically deforms ductile materials. The tool moves the base material, and depending on the geometric configuration of the tool, plastically deforms that material to the defined and repeatable shape. Only one tool is required for each desired configuration. The tool with one-point landing channels the material while deforming or lifting the material to the desired geometric shape and angle, simultaneously finning and forming microchannels of a variety of repeatable contours and dimensions.

1.2.1.2 Micro-Sawing

Micro-sawing is a technique widely used in industry that can fabricate rectangular channels in metal or silicon with an applicable channel width in the range of 0.1–10 mm. This technology can fabricate microchannels with high or low aspect ratios. It is very fast and has the lowest manufacturing cost among all micro-fabrication technologies. The technology uses a fret saw to fabricate rectangular microchannels, such as those shown in Fig. 1.5.

Fig. 1.4 Working principle of micro-deformation technology: *1* work piece, *2* cutting tool, *3* section with microchannels (Kukowski 2003)

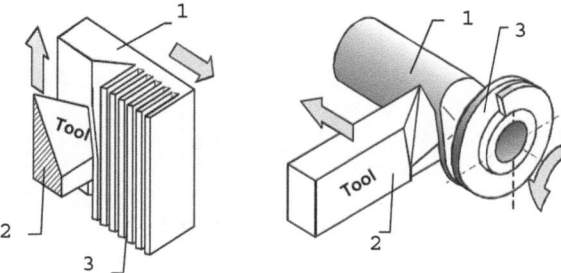

Fig. 1.5 Microchannels manufactured by micro-sawing technology (Jang et al. 2003)

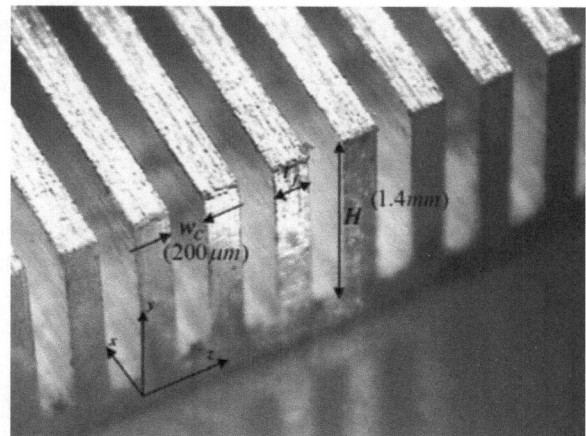

1.2.2 Modern Technology

1.2.2.1 MEMS (Micro-Electro-Mechanical Systems)

Much of the current research in MEMS centers on a group of batch micro-fabrication methods that arose from the semiconductor sector. Many technologies are included among MEMS methods: wet etching, dry etching, LIGA (Lithographie, Galvanoformung, and Abformung), and deep reactive ion etching (DRIE). This section will focus on DRIE technology, which is the most widely used among MEMS technologies. As shown in Table 1.1, rectangular, circular, triangular, or trapezoidal channels can be fabricated using the DRIE technique. This technology is applicable to metal, silicon, and glass with a wide range of channel sizes, from the nanometer scale to the millimeter scale. In addition, the technology has the advantage of low manufacturing uncertainty. However, the DRIE technology is not well suited for use in industrial fields due to its time-consuming process.

The DRIE technology process shown in Fig. 1.6 is as follows: (1) Deposit Photo Resistor (PR) material on the specimen using a sputtering process; (2)

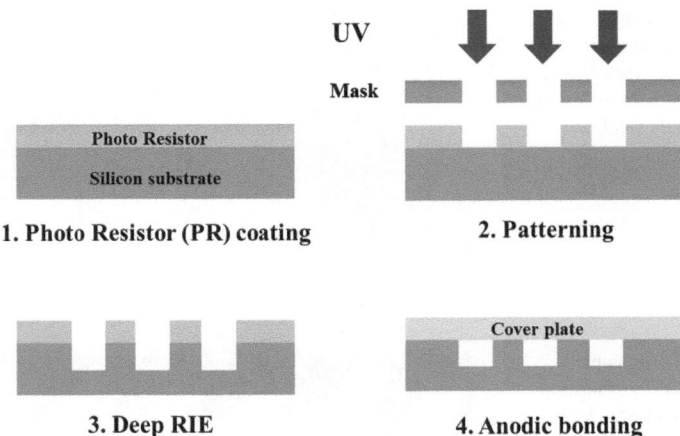

Fig. 1.6 Process of DRIE technology (Youn and Kim 2012)

Expose the specimen to UV light; (3) Form the channels by the DRIE process; (4) Join a cover with the specimen by anodic bonding.

1.2.2.2 Laser Micro-Machining

Recently, laser micro-machining technology has been applied to the fabrication of microchannels. Laser micro-machining is applicable to any material and can produce a wide range of channel sizes from the nanometer to millimeter scale in an unlimited number of geometries. In addition, the technology has the advantage of low manufacturing uncertainty. Laser micro-machining technology is better than the technologies discussed above in all aspects except cost and process speed. For this reason, the technique has not yet been adopted by industry. A diagram of a laser micro-machining system is shown in Fig. 1.7a. The laser enters a test section through a focusing lens while a computer-controlled XYZ-stage moves the specimen to form the microchannels. Figure 1.7b and c shows some microchannels manufactured by the laser micro-machining process.

1.3 Single-Phase Flow in Microchannels

Many experimental correlations for pressure drop have been developed from experimental measurements. Since the pioneering work done by Tuckerman and Pease (1981) for high heat flux removal by microchannel arrays, much of the research has concentrated on fluid flow paths having constant cross-section. In 1981, Tuckerman and Pease investigated the ability of microchannels to cool integrated circuits. They reported that water-cooled microchannels fabricated on

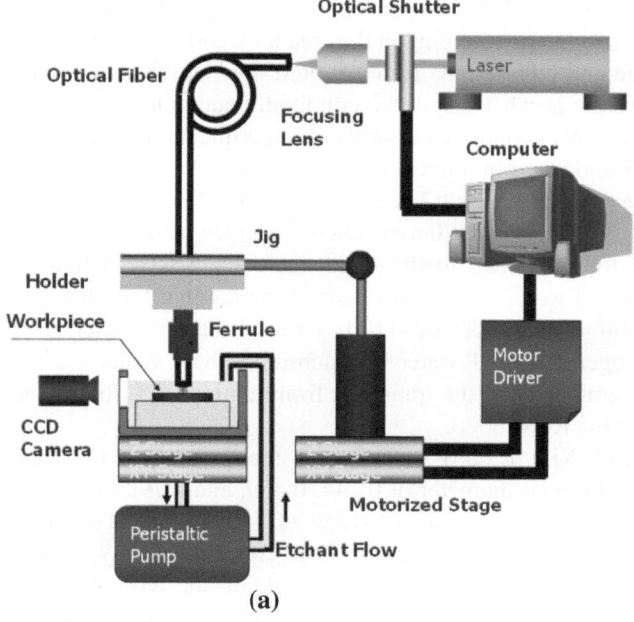

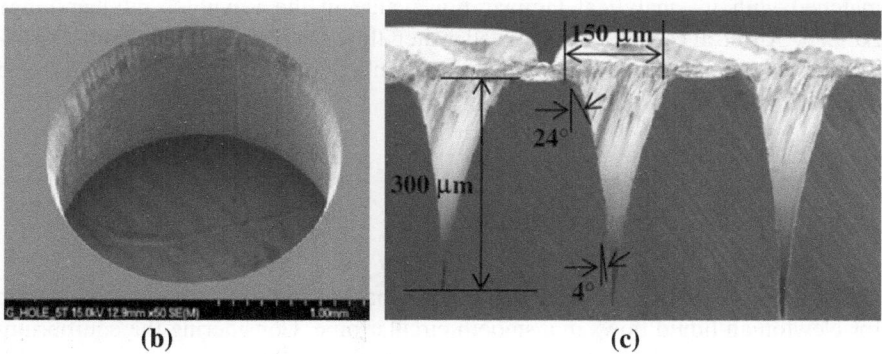

Fig. 1.7 a Schematic diagram of a laser micro-machining system; **b** Drilling of 1 μm-thick glass with hole diameter of 1 μm; **c** A scanning electron microscope image of the cross-section of microchannels fabricated using femtosecond laser micro-machining (Lim et al. 2008)

the substrate of the IC chip could dissipate heat of 790 W/cm^2 without a phase change and with maximum substrate temperature rise of 71 °C above the inlet water temperature. Shortly after Tuckerman and Pease, Wu and Little (1983, 1984) conducted several experiments with gas flowing instead of liquid in the trapezoidal-shaped silicon/glass microchannel to measure the flow friction and heat transfer characteristics. They reported that the transition from laminar to turbulent flow occurs at Reynolds numbers of 400–900 depending on test conditions. They suggested that reducing the transition Reynolds number improved the heat transfer.

Pfahler et al. (1990) experimentally investigated three microchannels of rectangular cross-section ranging in the area 80–7200 μm^2 using N-propanol as the working fluid and reported the result of fluid flow and friction factor. Their target was to determine at what length the continuum equation broke down. Later, they (Pfahler et al. 1991) continued a series of experiments to measure friction factors using liquids and gases in microchannels. Peng and Wang (1993), Wang and Peng (1994), Peng et al. (1995), and Peng and Peterson (1996) performed studies on heat transfer and fluid flow for different microchannnels and microchannel structures. They tested for water and methanol in microchannels with hydraulic diameters ranging from 0.3 to 0.75 mm and obtained the laminar/turbulent transition in the Reynolds number range of 700–1100. Yu et al. (1995) presented results for the flow of nitrogen gas and water in microtubes with diameters of 19, 52, and 102 μm. In laminar flow, the transition from laminar to turbulent occurred in the range of $2000 < Re < 6000$.

Hwang and Kim (2006) investigated the pressure drop characteristics in microtubes with inner diameter of 0.244, 0.430, and 0.792 mm using R-134a as a test fluid in the Reynolds number range of 150–10,000. Yen et al. (2003) performed an experimental investigation in microtubes of 0.19, 0.30, and 0.51 mm using HCFC123 and FC-72 as the working fluids and reported the results of fluid flow and heat transfer. They showed that the friction factor in microtubes is well matched with its analytical laminar flow value in the Reynolds number range 20–265. Xu et al. (2000) investigated water flow in microchannels both experimentally and analytically, with Reynolds numbers ranging from 20 to 4000 and hydraulic diameters from 30 to 344 μm.

1.3.1 Pressure Drop Correlations

The following equations are readily derived based on the continuum assumption for Newtonian liquid flows in a smooth circular pipe. Considering the equilibrium of a fluid element of length dx in a pipe of diameter d, the force due to pressure difference dp is balanced by the frictional force due to shear stress τ_w at the wall:

$$\left(\frac{\pi}{4}d^2\right)dp = (\pi d\, dx)\tau_w \qquad (1.1)$$

The pressure gradient and the wall shear stress are thus related by the following equation:

$$\frac{dp}{dx} = \frac{4\tau_w}{d} \qquad (1.2)$$

For Newtonian fluids, the wall shear stress τ_w is expressed in terms of the velocity gradient at the wall:

$$\tau_w = \mu \frac{du}{dy}\bigg|_w \tag{1.3}$$

where μ is the viscosity of fluid. The friction factor f is thus related by the following equation:

$$f = \frac{\tau_w}{(1/2)\rho u_m^2} \tag{1.4}$$

where u_m is the mean flow velocity in the channel.

The frictional pressure drop p over a length L is obtained from Eqs. (1.2) and (1.4):

$$\Delta p = \frac{2f\rho u_m^2 L}{d} \tag{1.5}$$

For non-circular flow channels, the d in Eq. (1.5) is replaced by the hydraulic diameter d_h represented by the following equation:

$$d_h = \frac{4A_c}{P_w} \tag{1.6}$$

where A_c is the flow-channel cross-sectional area and P_w is the wetted perimeter. For a rectangular channel of sides a and b, d_h is given by

$$d_h = \frac{4ab}{2(a+b)} \tag{1.7}$$

The friction factor f in Eq. (1.5) depends on the flow conditions. The following relation for the friction factor of a laminar flow is theoretically given as

$$f = \frac{64}{\mathrm{Re}} \tag{1.8}$$

where the constant 64 is changed by the geometry of the channel cross-section. Table 1.2 shows the list of values of the constant (called the Poiseuille number) for different geometries. As shown in Fig. 1.8, Eq. (1.8) is well matched with the experimental data of Yen et al. (2003) for the microchannels of 0.19 and 0.51 mm.

For turbulent flows many types of relations are available in the literature for friction factor. These are generally defined by the Blasius equation as

$$f = 0.316\mathrm{Re}^{-1/4} \tag{1.9}$$

As shown in Fig. 1.9, Eq. (1.9) is well matched with the experimental data of Hwang and Kim (2006) for the microchannels of 0.244, 0.43, and 0.792 mm.

Table 1.2 Characteristic values of laminar flow in circular and noncircular channels (Celata 2004)

Channel cross section	Channel geometry	Hydraulic diameter	Constant
Circle	diameter d	d_h	64
Rectangular	a, b, a/b = 0.1	2ab/(a + b)	85.76
Rectangular	a, b, a/b = 0.2	2ab/(a + b)	76.8
Rectangular	a, b, a/b = 0.4	2ab/(a + b)	65.28
Rectangular	a, b, a/b = 0.6	2ab/(a + b)	60.16
Rectangular	a, b, a/b = 0.8	2ab/(a + b)	57.6
Square	Side a	a	56.96

Fig. 1.8 Comparison between the frictional factor of turbulent flow and the experimental results of Yen et al. (2003)

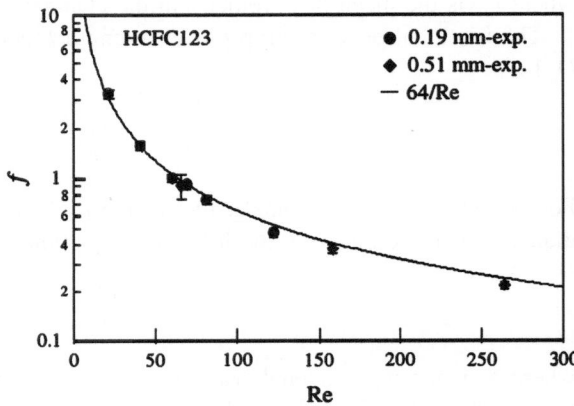

Fig. 1.9 Comparison between the frictional factor of turbulent flow and the experimental results of Hwang and Kim (2006)

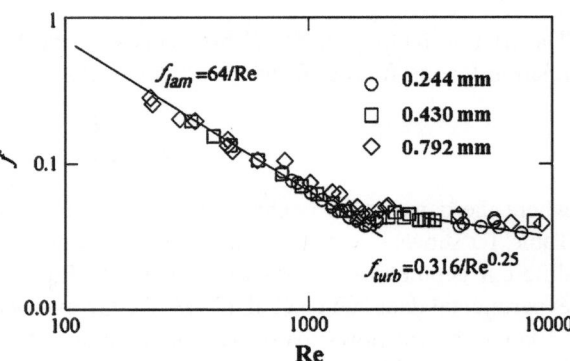

Fig. 1.10 Nusselt number as a function of dimensionless length and experimental results compared with Eq. (1.10) (Shilder et al. 2010)

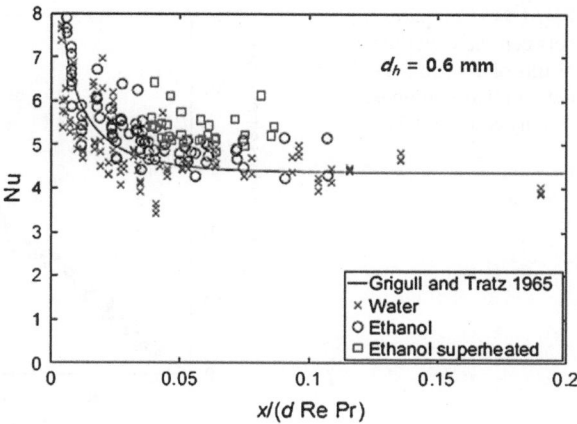

1.3.2 Heat Transfer Correlations

1.3.2.1 Laminar Flow

The Nusselt number of a fully developed laminar flow is 4.36 when there is a constant heat flux boundary condition at the tube wall is a constant heat flux. Grigull and Tratz (1965) numerically investigated the thermal entrance problem for laminar flow with constant heat flux. They evaluated the Nusselt number as a function of the dimensionless axial distance, Reynolds number, and Prandtl number:

$$Nu = 4.36 + \frac{0.00668(d_h/x)\text{Re Pr}}{1 + 0.04[(d_h/x)\text{Re Pr}]^{2/3}} \qquad (1.10)$$

Shilder et al. (2010) conducted an experiment for single phase flow in a microchannel that had a hydraulic diameter of 0.6 mm. If the experimental uncertainties are accounted for, as shown in Fig. 1.10, the measurements by Shilder et al. (2010) agree with the classical theory of Eq. (1.10).

1.3.2.2 Turbulent Flow

Adams et al. (1997) conducted experimental work in the turbulent region with water flow in circular microchannels of 0.76 and 0.109 mm. Based on their data, they proposed the following equation:

$$\text{Nu} = \text{Nu}_{Gn}(1 + F) \qquad (1.11)$$

Fig. 1.11 Comparison between the experimental results of Yu et al. (1995) and the correlation proposed by Adams et al. (1997)

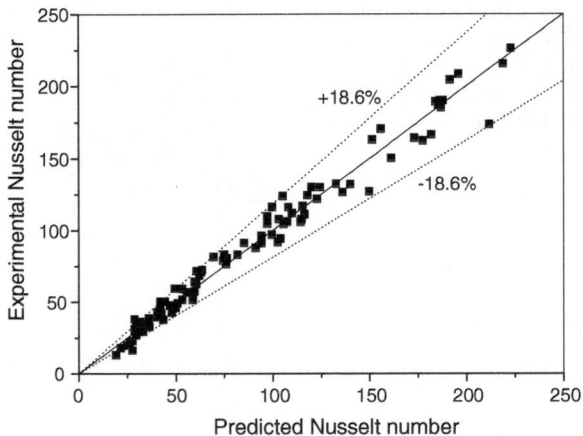

where

$$\mathrm{Nu}_{Gn} = \frac{(f/8)(\mathrm{Re} - 1000)\,\mathrm{Pr}}{1 + 12.7(f/8)^{1/2}(\mathrm{Pr}^{2/3} - 1)} \tag{1.12}$$

$$f = (1.82\log(\mathrm{Re}) - 1.64)^{-2} \tag{1.13}$$

$$F = 7.6 \times 10^{-5}\mathrm{Re}\left(1 - (d_h/d_o)^2\right) \tag{1.14}$$

Nu_{Gn} represents the Nusselt number predicted by Gnielinski's (1976) correlation. The least-squares fit to all the data sets studied by Adams et al. (1997) resulted in $d_o = 1.164$ mm. As shown in Fig. 1.11, the correlations are well matched with the experimental data for the 0.102, 0.76, and 1.09 mm microchannels of Yu et al. (1995), within ±18.6 %.

1.4 Phase-Change (multi-phase) Flow in Microchannels

1.4.1 Pressure Drop Correlations

Many investigations have examined the pressure drop correlations for two-phase flow (Ghiaasiaan, 2008). Lockhart and Martinelli (1949) defined a two-phase friction multiplier to relate the two-phase pressure gradient to the single-phase pressure gradient for liquid flow:

$$\left(\frac{\mathrm{d}P}{\mathrm{d}z}\right)_{TP} = \left(\frac{\mathrm{d}P}{\mathrm{d}z}\right)_{L}\Phi_L^2 \tag{1.15}$$

Table 1.3 Constant and exponents in the correlation of Lee and Lee (2001)

Liquid regime	Gas flow regime	A	q	r	s
Laminar	Laminar	6.833×10^{-8}	-1.317	0.719	0.577
Laminar	Turbulent	6.185×10^{-2}	0	0	0.726
Turbulent	Laminar	3.627	0	0	0.174
Turbulent	Turbulent	0.408	0	0	0.451

The friction multiplier, Φ_L^2, is defined by coefficient C and the Lockhart-Martinelli parameter, X^2, as the ratio of the single-phase liquid and gas pressure gradients.

$$\Phi_L^2 = 1 + \frac{C}{X} + \frac{1}{X^2} \tag{1.16}$$

$$X = \sqrt{\frac{(dP/dz)_L}{(dP/dz)_G}} \tag{1.17}$$

Chisholm and Laird (1958) related the friction multiplier to the Lockhart-Martinelli parameter through a simple expression that depends on the coefficient C ranging from 5 to 20, depending on laminar or turbulent flow of gas and liquid. Some researchers have suggested empirical correlations for the coefficient C to determine the two-phase friction multiplier. Among the suggested correlations, the most widely used correlations are Mishima and Hibiki's correlation (1996) and Lee and Lee's correlation (2001). Mishima and Hibiki's correlation appears to provide the best correlation for adiabatic two-phase flow, although its applicability to minichannel flows with phase change has not been demonstrated. They have proposed

$$C = 21(1 - e^{-0.319d_h}) \tag{1.18}$$

where the diameter d_h is in millimeters. Cavallini et al. (2005) have recently shown that Mishima and Hibiki's method can predict the two-phase pressure drop for flow condensation of refrigerants R-134a and R-236ea in 1.4 mm tubes. The correlation of Mishima and Hibiki (1996) evidently assumes that C depends on channel size only. Based on the observation that C depends on phase mass fluxes as well, and using experimental data from several sources as well as their own data that covered channel gaps in the 0.4 to 4 mm range, Lee and Lee (2001) derived the following correlation for C, for adiabatic flow in horizontal thin rectangular channels:

$$C = A\left(\frac{\mu_L^2}{\rho_L \sigma d_h}\right)^q \left(\frac{\mu_L j}{\sigma}\right)^r \operatorname{Re}_{L0}^s \tag{1.19}$$

where j represents the total mixture volumetric flux. The constants A, r, q, and s depend on the liquid and gas flow regimes (viscous or turbulent), and their values are listed in Table 1.3.

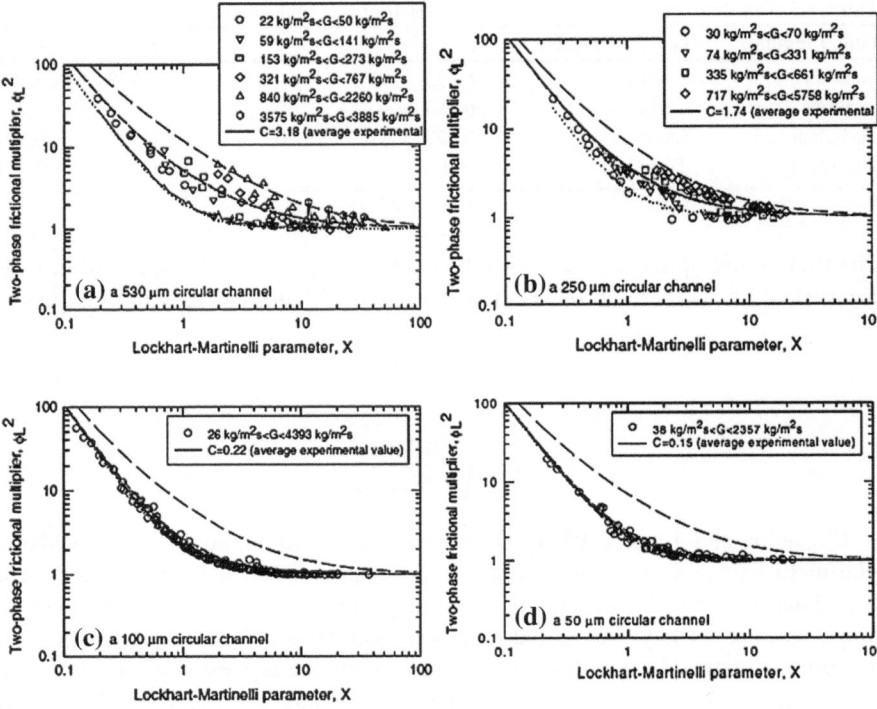

(– – –) Chisholm (1967), (- - -) Mishima and Hibiki (1996), (——) Lee and Lee (2001)

Fig. 1.12 Pressure drop characteristics of two-phase flow: Variation of the two-phase multiplier with the Lockhart-Martinelli parameter (Chung and Kawaji 2004)

The correlations of Mishima and Hibiki (1996) and Lee and Lee (2001), Eqs. (1.18) and (1.19), respectively, predicted the data of Chung et al. (2004) for adiabatic flow of water and nitrogen in horizontal 96 μm square rectangular microchannels, the data of Zhao and Bi (2001) for water and air flow in a miniature triangular channel with $d_h = 0.87$–2.89 mm, and the data of Chung and Kawaji (2004) for water and nitrogen flow in a horizontal circular channel with $d_h = 50$–530 μm, within about 10 % accuracy. Figure 1.12 shows the two-phase friction multiplier data plotted against the Lockhart-Martinelli parameter for the data of Chung and Kawaji (2004).

1.4.2 Heat Transfer Correlations

A sequence of two-phase and boiling heat transfer regimes takes place along the heated channels during flow boiling. The sequence of flow regimes includes bubbly, slug, churn, and annular flow. Following the initiation of boiling, the sequence of heat transfer regimes includes subcooled boiling, saturated nucleate boiling, saturated forced-flow boiling, and dry out field. Among the heat transfer regimes observed in flow boiling, the most commonly observed regime is the saturated flow boiling regime, which will be the focus of this section. Saturated, forced-flow boiling refers to the entire region between the point where $x = 0$ and the critical heat flux point. A sequence of complicated two-phase flow patterns, including bubbly, churn, slug, and annular-dispersed, can take place. The two-phase flow regimes cover a quality range of a few percent, up to very high values characteristic of annular flow regime (sometimes approaching 100 %). Nucleate boiling is predominant where quality is low (a few percent); forced convective evaporation is predominant at high qualities representing annular flow; and elsewhere both mechanisms can be important. The relative contribution of forced convection increases as quality increases.

Nucleate boiling and forced convective evaporation both contribute to the heat transfer in saturated flow boiling. At low x, the contribution of the nucleate boiling mechanism dominates, but the contribution of convection increases as quality is increased. Once the annular-dispersed flow regime is achieved, the contribution of convective evaporation becomes predominant. Forced-flow boiling correlations should thus take into account the composite nature of the boiling heat transfer mechanism. Correlations for saturated flow boiling are reviewed next.

1.4.2.1 Chen's Correlation (1966)

Chen's correlation is among the oldest, most successful and widely used correlations for saturated boiling. It works well for water at relatively low pressure and has been applied to a variety of fluids. The correlation can be used for $d_h \geq 1$ mm, $P = 0.09\text{--}3.45$ Mpa, $x = 0\text{--}0.7$, $q'' = 0\text{--}2.4$ MW/m^2.

$$h = h_{NB} + h_{FC} \qquad (1.20)$$

The forced convection component is found from

$$h_{FC} = 0.023\text{Re}_f^{0.8}\text{Pr}_f^{0.4}F\left(\frac{k_f}{d_h}\right) \qquad (1.21)$$

where

$$\text{Re}_f = G(1 - x)d_h/\mu_f \qquad (1.22)$$

$$\text{Pr}_f = (\mu C_p/k)_f \qquad (1.23)$$

The factor F represents $(\text{Re}_{TP}/\text{Re}_f)^{0.8}$ and was correlated by Chen empirically in a graphical form. A curve fit to the graphical correlation is (Collier 1981):

$$F = \begin{cases} 1 & \text{for } X_u^{-1} < 0.1 \\ 2.35(0.213 + X_u^{-1})^{0.736} & \text{for } X_u^{-1} > 0.1 \end{cases} \quad (1.24)$$

The nucleate boiling component is based on the correlation of Forster and Zuber (1955), modified to account for the reduced average superheat in the thermal boundary layer on bubble nucleation on wall cavities:

$$h_{NB} = 0.00122 \left\{ \frac{k_f^{0.79} C_{pf}^{0.45} \rho_f^{0.49} g_c^{0.43}}{\sigma^{0.5} \mu_f^{0.29} h_{fg}^{0.24} \rho_g^{0.24}} \right\} \Delta T_{sat}^{0.24} \Delta P_{sat}^{0.75} S \quad (1.25)$$

where $\Delta T_{sat} = T_w - T_{sat}$ and $\Delta P_{sat} = P_{sat}(T_W) - P$. Note that g_c is needed for English units only. The parameter S is Chen's suppression factor and represents $S = (\Delta T_{eff}/\Delta T_{sat})^{0.99}$, where ΔT_{eff} is the effective liquid superheat in the thermal boundary layer. S was also correlated graphically. An empirical curve fit to Chen's graphical correlation is (Collier 1981):

$$S = \left[1 + (2.56 \times 10^{-6})(\text{Re}_f F^{1.25})^{1.17} \right]^{-1} \quad (1.26)$$

1.4.2.2 Kandlikar's Correlation (1990)

Kandlikar's correlation is based on 10,000 data points for water, refrigerants and cryogenic fluids. The correlation can be used for $d_h \geq 1$ mm.

$$h = \max(h_{NBD}, h_{CBD}) \quad (1.27)$$

$$h_{CBD} = \left\{ 1.136\text{Co}^{-0.9}(1 - x)^{0.8} f_2(\text{Fr}_{f0}) + 667.2\text{Bl}^{0.7}(1 - x)^{0.8} F_{fl} \right\} h_{f0} \quad (1.28)$$

$$h_{NBD} = \left\{ 0.6683\text{Co}^{-0.2}(1 - x)^{0.8} f_2(\text{Fr}_{f0}) + 1058\text{Bl}^{0.7}(1 - x)^{0.8} F_{fl} \right\} h_{f0} \quad (1.29)$$

where

$$h_{f0} = \frac{k_f}{d_h} \frac{(\text{Re}_f - 1,000)(f/2)\,\text{Pr}_f}{[1 + 12.7(\text{Pr}_f^{2/3} - 1)(f/2)^{0.5}]} \quad (1.30)$$

The parameters in Kandlikar's correlation are the convection number Co, the boiling number Bl, and the Froude number when all mixture is saturated liquid, Fr_{f0}, defined, respectively, as

$$\text{Co} = (\rho_g/\rho_f)^{0.5}[(1 - x)/x]^{0.8} \quad (1.31)$$

$$\text{Bl} = q_w''/(Gh_{fg}) \quad (1.32)$$

Table 1.4 Values of the fluid-surface parameter F_{fl} in the correlation of Kandlikar (1997)

Fluid	F_{fl}	Fluid	F_{fl}
Water	1.00	R-32/R-1132	3.30
R-11	1.30	R-124	1.00
R-12	1.50	R-141b	1.80
R-13BI	1.31	R-134a	1.63
R-22	2.20	R-152a	1.10
R-113	1.30	Kerosene	0.488
R-114	1.24	Nitrogen	4.70
		Neon	3.50

Note Use 1.0 for any fluid with a stainless steel tube

and

$$\mathrm{Fr}_{f0} = G^2/(\rho_f^2 g d_h) \tag{1.33}$$

The parameter F_{fl} is the fluid-surface parameter discussed above (see Table 1.4). Finally,

$$f_2(\mathrm{Fr}_{f0}) = \begin{cases} 1 & \text{for } \mathrm{Fr}_{f0} \geq 0.4 \\ (25\mathrm{Fr}_{f0})^{0.3} & \text{for } \mathrm{Fr}_{f0} < 0.4 \end{cases} \tag{1.34}$$

1.4.2.3 Gungor and Winterton's Correlation (1986)

Gungor and Winterton's correlation is based on 3,700 data points for water, refrigerants, and ethylene glycol. The correlation can be used for $d_h \geq 1$ mm. The original correlation (Gungor and Winterton 1986) was subsequently simplified by the authors (Gungor and Winterton 1987) to the following easy-to-use correlation:

$$h = h_f\left\{1 + 3000\mathrm{Bl}^{0.86} + 1.12[x/(1-x)]^{0.75}[\rho_f/\rho_g]^{0.41}\right\}E_2 \tag{1.35}$$

where

$$E_2 = \begin{cases} 1 & \text{for } \mathrm{Fr}_{f0} \geq 0.05 \\ \mathrm{Fr}_{f0}^{(0.1-2\mathrm{Fr}_{f0})} & \text{for } \mathrm{Fr}_{f0} < 0.05 \end{cases} \tag{1.36}$$

1.4.2.4 Shah's Correlation (1982)

Shah (1982) proposed correlations to implement his chart calculation method and recent comparisons show his correlations perform satisfactorily for mini- and macrochanenls. Shah's correlations consider nucleate and convective boiling both to be important to the two-phase flow evaporative heat transfer. Similar to Chen

(1966), his method chooses the larger of the two as the dominant contributor to the local two-phase flow boiling heat transfer coefficient. Shah's method is meant to be applicable to both vertical and horizontal tubes.

When $N > 1.0$ and $Bl > 0.0003$, h is calculated as below:

$$h = 230Bl^{0.5}h_f \tag{1.37}$$

where

$$N = \left(\frac{1-x}{x}\right)^{0.8}\left(\frac{\rho_g}{\rho_f}\right)^{0.5} \tag{1.38}$$

When $N > 1.0$ and $Bl < 0.0003$, h is calculated as below:

$$h = \left(1 + 46Bl^{0.5}\right)h_f \tag{1.39}$$

When $1.0 > N > 0.1$, h is calculated as below:

$$h = F_sBl^{0.5}\exp(2.74N - 0.1)h_f \tag{1.40}$$

When $N < 0.1$, h in the bubble suppression regime is calculated using the equation below:

$$h = F_sBl^{0.5}\exp(2.74N - 0.15)h_f \tag{1.41}$$

In the above equations, Shah's constant $F_s = 14.7$ when $Bl > 0.0011$ and $F_s = 15.43$ when $Bl < 0.0011$. Shah (2006) compared several correlations for conventional channels against a wide range of data that included 30 pure fluids. Best results were found with the correlations of Shah (1982) and Gungor and Winterton (1987), the mean deviation for all data being about 17 %.

1.4.2.5 Li and Wu's Correlation (2010)

Recently, Li and Wu obtained a correlation using the boiling number, Bond number, and Reynolds number. The correlation contains more than 3,744 data points, covering a wide range of working fluids, operational conditions, and different microchannel dimensions. In addition, they showed that the Bond number in predicting heat transfer coefficients can be used as a criterion to classify a flow path as a microchannel or as a conventional macrochannel. The correlation can be used for 0.19 mm $\leq d_h \leq 2.01$ mm.

$$h = 334Bl^{0.3}(BoRe_f^{0.36})^{0.4}(k_f/d_h) \tag{1.42}$$

where

$$Bl = q_w''/(Gh_{fg}) \tag{1.43}$$

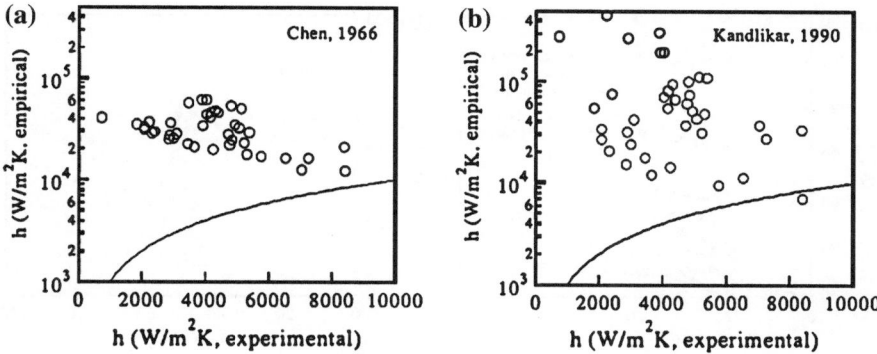

Fig. 1.13 Comparison between the experimental results of Yen et al. (2003) and **a** the correlation proposed by Chen (1996), and **b** the correlation proposed by Kandlikar (1990)

$$\text{Bo} = \frac{g(\rho_l - \rho_g)d_h^2}{\sigma} \tag{1.44}$$

1.5 Comparison of the Selected Correlations

The correlations by Chen, Kandlikar, and Gungor and Winterton predict experimental data well for cases where the channel hydraulic diameter is larger than 0.5 mm. However, recent studies by Yen et al. (2003) and Li and Wu (2010) show that the correlations by Chen, Kandlikar, and Gungor and Winterton over-predict or under-predict the experimental data of microchannels. As shown in Fig. 1.13, Chen's correlation and Kandikar's correlation under-predicted the experimental data of Yen et al. (2003) by more than an order of magnitude. In addition, as shown in Fig. 1.14, Gungor and Winterton's correlation over-predicted the experimental data for channels having hydraulic diameters of 0.586 and 0.19 mm, while the correlation was well matched with the data for the 2.01 mm channel. However, Li and Wu's correlation predicted the experimental data well for the range of hydraulic diameters from 0.19 to 2.01 mm.

1.5.1 Two-Phase Flow Correlations Without Phase Change

Gas–liquid flows have been studied extensively due to their presence in nature and usefulness in many important industrial applications including oil transport, steam generation, bubble columns, reactors, aeration systems for chemical processes, and cooling systems for energy production. More recently, two-phase flows within microchannels have been receiving attention because of their applicability to such fields as MEMS, electronics cooling, medical and genetic engineering, and

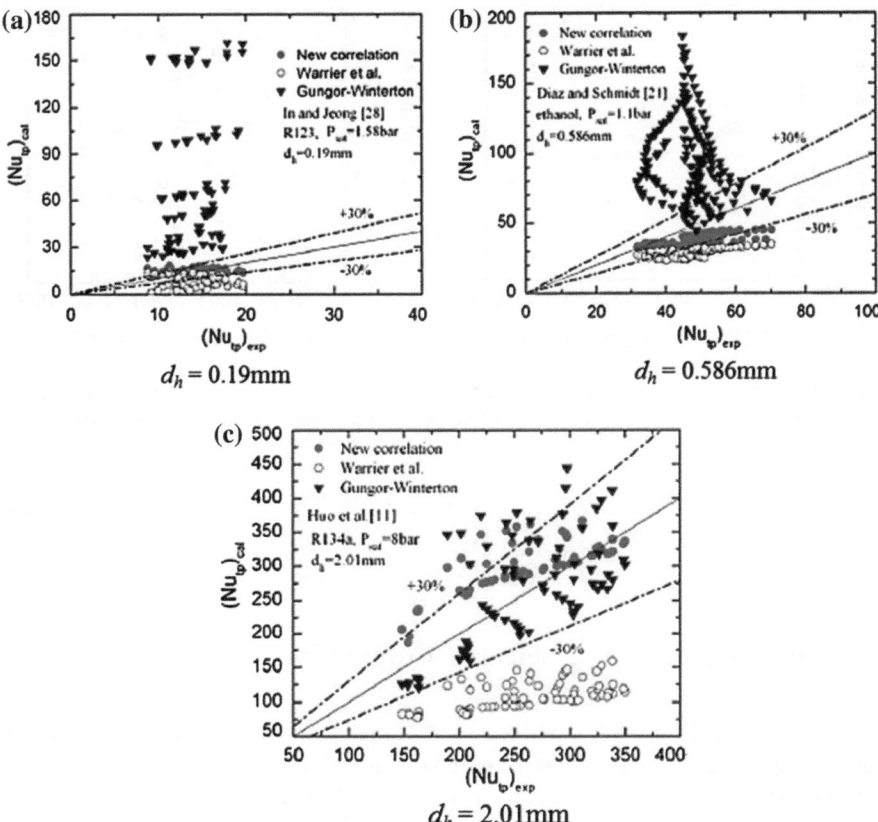

Fig. 1.14 Comparison between the experimental results and the correlations: **a** $d_h = 0.19$ mm; **b** $d_h = 0.586$ mm; and **c** $d_h = 2.01$ mm (Li and Wu 2011)

bioengineering. To design and evaluate the performance of devices in these fields, investigators must have fundamental knowledge of two-phase flow characteristics in small flow passages, including flow pattern, void fraction, pressure drop, and heat transfer coefficient.

Many studies have been conducted on the flow and heat transfer characteristics of two-phase flows within microchannels because of their practical applications. The majority of research on two-phase flow has focused on configurations in which flow boiling arises, and thus many flow boiling heat transfer characteristics and empirical correlations have been presented (Schilder et al. 2010; Morini 2004; Lelea et al. 2004; Celata et al. 1993; Choo and Kim 2010a). In addition, many researchers have observed flow patterns in microchannels and presented flow pattern maps for the purpose of understanding the relationship between heat transfer and flow characteristics (Serizawa et al. 2002; Chung and Kawaji 2004; Triplett et al. 1999).

Fig. 1.15 A plot of the experimentally determined heat transfer coefficient as a function of the superficial gas velocity and the gas Reynolds number (Hetsroni et al. 2009)

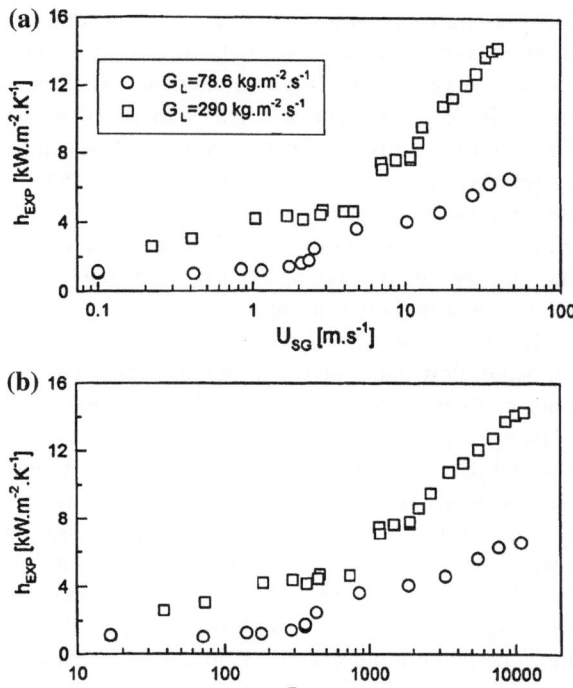

Recently, heat transfer experiments were conducted by Bao et al. (2000) using a 1.95 mm diameter minichannel under non-boiling, air–water flow conditions. The results of this research showed that heat transfer performance increased with increasing air flow rate at a fixed liquid flow rate. Flow patterns were not observed as a part of this work, but the investigators suggested that the augmentation in heat transfer was caused by the transition of the flow regime from bubbly to intermittent slug to annular. Hetsroni et al. (2009) conducted research using a 0.1 mm microchannel using an air–water flow under non-boiling conditions. The microchannel was manufactured onto a silicon substrate using a MEMS process. They observed decreasing heat transfer performance as the air flow rate was increased under fixed water flow rate conditions. The heat transfer results for microchannels of Hetsroni et al. (2009) showed opposite trends from those for minichannels of Bao et al. (2000). Therefore, it is expected that there exists a transition diameter that distinguishes the heat transfer characteristics of microchannels from minichannels. Finally, Choo and Kim (2011) showed the transition diameter that distinguishes the heat transfer characteristics of microchannels from minichannels.

Bao et al. (2000) studied a 1.95 mm diameter minichannel under non-boiling, air–water flow conditions. Figure 1.15 shows the experimental heat transfer coefficients for horizontal flows with (a) a liquid mass flux of 78.6 kg m^{-2}s^{-1} and a heat flux of 20 kW m^{-2}, and (b) a liquid mass flux of 290 kg m^{-2}s^{-1} and a heat flux of 33 kW m^{-2} plotted against the gas superficial velocity and plotted against the gas-phase Reynolds number at the heating section. As expected, the heat transfer coefficient increases with increasing liquid mass flux and increasing gas superficial velocity or Reynolds number. The data show a sharp increase in the heat transfer coefficient at a gas velocity of about 3 m/s and a gas superficial velocity of 5.5 ms^{-1}. The measured heat transfer coefficients for the air–water system are always higher than would be expected for the corresponding single-phase liquid flow, so the addition of air can be considered to have an enhancing effect. This effect presumably arises from the increased liquid velocity caused by the presence of the gas phase and from higher turbulence intensities due to the relative motion between the phases.

Hetsroni et al. (2009) conducted research using a 0.1 mm microchannel using air–water flow under non-boiling conditions. The microchannel was manufactured onto a silicon substrate using a MEMS process. With increasing superficial gas velocity, a gas core with a thin liquid film was observed. The visual observation showed that as the air velocity increased, the liquid droplets entrained in the gas core disappeared such that the flow became annular. The probability of appearance of different flow patterns should be taken into account for developing flow pattern maps. The dependence of the Nusselt number on liquid and gas Reynolds numbers, based on liquid and gas superficial velocity, respectively, was determined to be in the range of $Re_{LS} = 4$–56 and $Re_{GS} = 4.7$–270. It was shown that an increase in the superficial liquid velocity involves an increase in heat transfer. This effect was reduced with increasing superficial gas velocity, in contrast to the results reported on two-phase heat transfer in conventional size channels.

The two-phase flow was achieved by introducing water and air into a mixer as shown in Fig. 1.16. The experiments were performed in an open loop, and therefore the outlet pressure was close to atmospheric. Two types of pumps were used: a peristaltic pump and a mini gear pump. The test module is shown in Fig. 1.17. It was fabricated by a 15 × 15 mm square-shaped silicon substrate with a thickness of 530 μm, which was covered by a Pyrex cover, 500 μm-thick, which served as both an insulator and a transparent cover through which flow in the microchannels could be observed. The Pyrex cover was anodically bonded to the silicon chip to seal the channels. In the silicon substrate, parallel microchannels were etched, the cross-section of each channel being an isosceles triangle. The angles at the base were 55°. A test module having 21 microchannels with hydraulic diameter of 130 μm was used.

A microscope with an additional camera joint was assembled to connect a high-speed camera to the microscope. A high-speed camera with a maximum frame rate of 10,000 fps was used to visualize the two-phase flow regimes in the micro-channels. In the parallel channels having common inlet and outlet collectors, non-uniform distribution of the working fluid occurred. Simultaneous different flow patterns were observed in different parallel microchannels, as were alternate flow patterns in a given microchannel. These patterns are illustrated in Fig. 1.18.

Results presented in Fig. 1.19 show the dependence of the Nusselt number, Nu_L, based on liquid thermal conductivity, and Reynolds number, Re_{GS}, based on superficial gas velocity and kinematic gas viscosity. As shown in Fig. 1.19, an

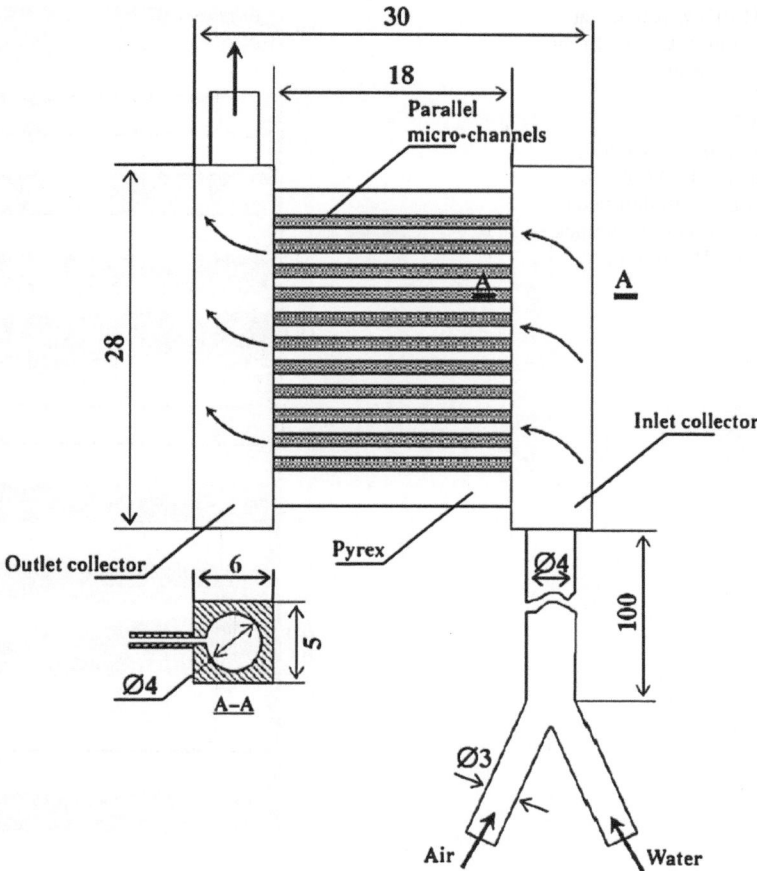

Fig. 1.16 Experimental facility. All dimensions in mm (Hetsroni et al. 2009)

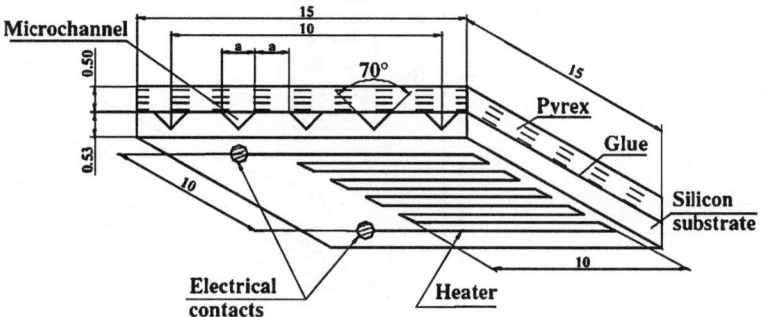

Fig. 1.17 Test module. All dimensions in mm (Hetsroni et al. 2009)

Fig. 1.18 Flow patterns at different times. $U_{GS} = 15$ m/s, $U_{LS} = 0.15$ m/s.
a t = 0.7320 s,
b t = 0.7770 s,
c t = 0.7940 s. L single-phase liquid, B bubbly flow, A1 gas core with thin liquid film; A2, gas core with thick liquid film (Hetsroni et al. 2009)

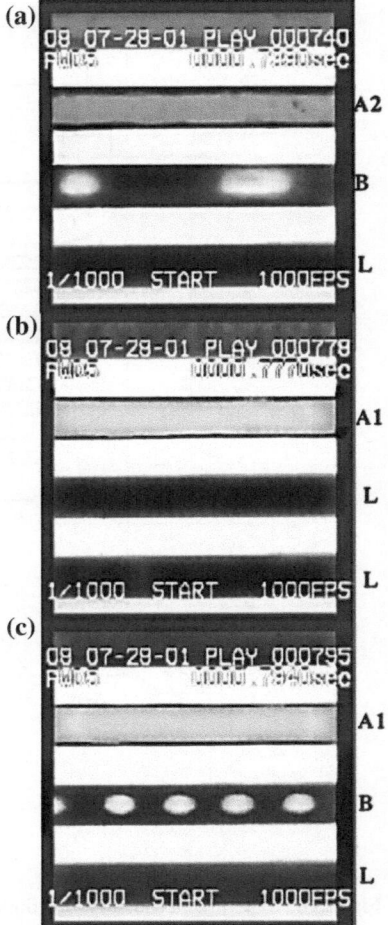

Fig. 1.19 Effect of superficial liquid velocity on heat transfer in parallel triangular microchannels of $d_h = 130$ μm (Hetsroni et al. 2009)

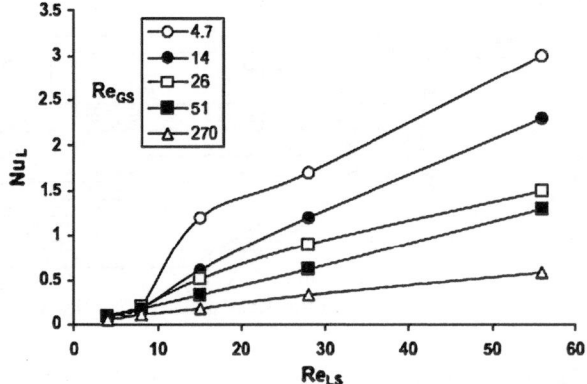

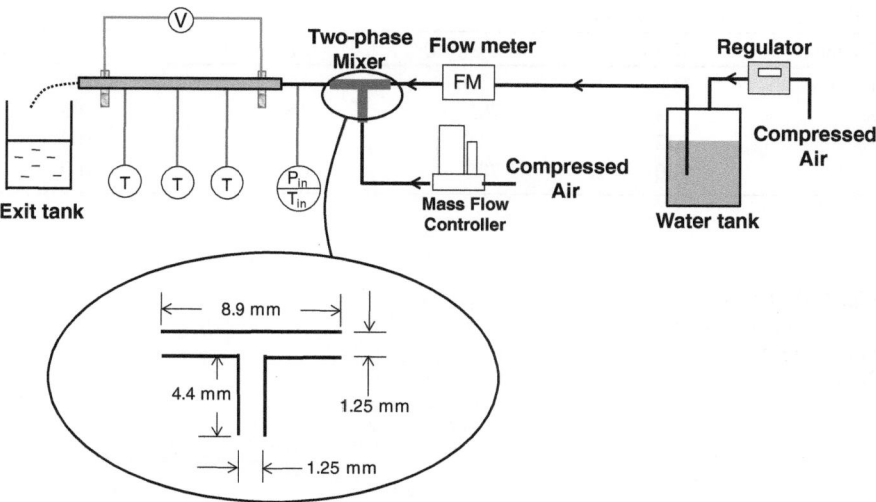

Fig. 1.20 Schematic diagram of experimental setup (Choo and Kim 2011)

increase in superficial liquid velocity leads to an increase in heat transfer. This effect falls off with increasing superficial gas velocity in the range $\text{Re}_{GS} = 4.7\text{--}270$. Based on the experimental data, the Nusselt number correlations were developed as shown in Eqs. (1.45) and (1.46).

$$\text{Nu}_L = 0.044\text{Re}_{LS}^{0.96}\text{Re}_{GS}^{-0.18} \quad \text{for} \quad \text{Re}_{GS} = 4.7 - 270, \text{Re}_{LS} = 4.0 - 8.0 \quad (1.45)$$

$$\text{Nu}_L = 0.13Re_{LS}^{0.96}\text{Re}_{GS}^{-0.40} \quad \text{for} \quad \text{Re}_{GS} = 4.7 - 270, \text{Re}_{LS} = 8.0 - 56 \quad (1.46)$$

Recently, Choo and Kim (2011) obtained the transition diameter that distinguishes the heat transfer characteristics of microchannels from minichannels. They experimentally investigated heat transfer and fluid flow characteristics of non-boiling two-phase flow in microchannels. The effects of channel diameter (140, 222, 334, and 506 μm) on the Nusselt number and the pressure drop were considered. Air and water were used as the test fluids. Results were presented for the Nusselt number and the pressure drop over a wide range of gas superficial velocity (1.24–40.1 m/s), liquid superficial velocity (0.57–2.13 m/s), and wall heat flux (0.34–0.95 MW/m²). The results showed that the Nusselt number increased with increasing gas flow rate for the large channels of 506 and 334 μm, while the Nusselt number decreased with increasing gas flow for the small channels of 222 and 140 μm. Based on these experimental results, a new correlation for the forced convection Nusselt number was developed. In addition, the two-phase friction multiplier decreased as channel diameter decreased due to the influence of viscous and surface tension forces.

Figure 1.20 shows a schematic diagram of the experimental apparatus. Four circular, stainless steel microchannels with inner diameters of 140, 222, 334, and 506 μm were used in the experiment. The channels were made of stainless steel

Table 1.5 Channel specifications

Channel	D_i (μm)	D_o (μm)	L_{heat} (mm)	L_{total} (mm)
1	506	1580	166	200
2	334	785	150	200
3	222	785	64	100
4	140	785	65	100

Fig. 1.21 Two-phase Nusselt numbers for each microchannel (Choo and Kim 2011)

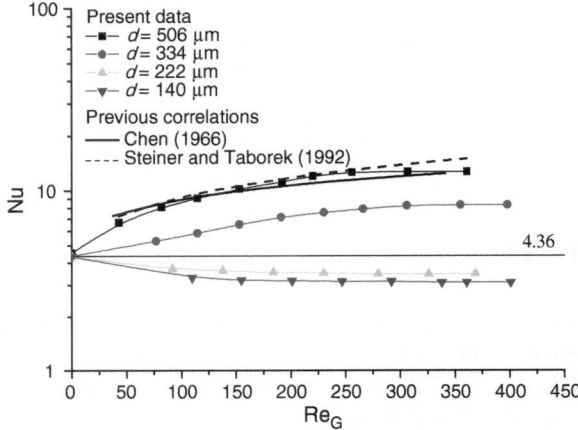

304, and the experimental conditions are listed in Table 1.5. Long channels having large length to diameter ratios ($395 \leq L/d \leq 714$) were used to diminish entrance and exit effects. The inner diameter of each microchannel was measured using a scanning electron microscope (SEM). A two-phase mixer was created using a T-junction to mix air and water. At the entrance and exit of the test section, similar T sections were used when measuring pressure and temperature.

The heat transfer characteristics for various air flow rates, at a fixed water flow rate condition, are shown in Fig. 1.21. For the 506 and 334 μm channels, as air flow rate was increased, heat transfer performance increased. This agrees with the results of Bao et al. (2000): the air injected into the minichannel causes turbulent mixing in the liquid film covering the wall. Thus, as the air flow rate is increased, the heat transfer performance increases. This was confirmed through flow pattern observation, as shown in Figs. 1.22 and 1.23. In order to observe the flow patterns, borosilicate glass capillary channels (145, 190, 303, and 506 μm) were used. As shown in Figs. 1.22a and 1.23a, slug and churn flow were observed at low Re_G. Liquid ring flow of churn flow appeared as air flow rate, and thus Re_G increased, as shown in Figs. 1.22b, c and 1.23b, c. A wavy interface was observed between the gas core and liquid film covering the channel wall, caused by turbulence. This agrees with previous observations by Serizawa et al. (2002).

On the other hand, for the 222 and 140 μm channels, at a fixed water flow rate, the heat transfer performance decreased as the air flow rate increased, as shown in

Fig. 1.22 Two-phase flow patterns within 506 μm channel for $Re_L = 420$: **a** $Re_G = 57$; **b** $Re_G = 197$; **c** $Re_G = 370$ V (Choo and Kim 2011)

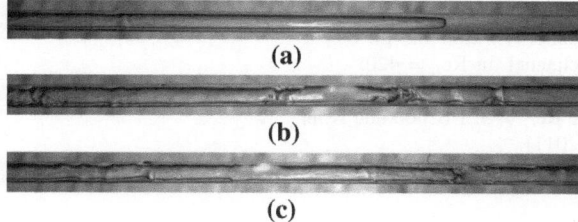

Fig. 1.23 Two-phase flow patterns within 303 μm channel for $Re_L = 420$: **a** $Re_G = 49$; **b** $Re_G = 176$; **c** $Re_G = 354$ (Choo and Kim 2011)

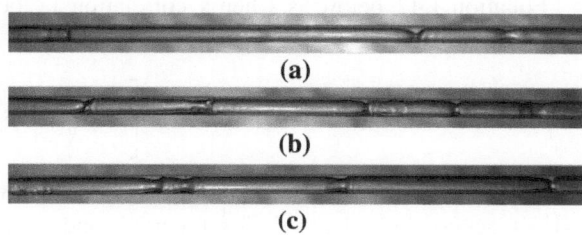

Fig. 1.24 Two-phase flow patterns within 190 μm channel for $Re_L = 420$: **a** $Re_G = 71$; **b** $Re_G = 152$; **c** $Re_G = 333$ (Choo and Kim 2011)

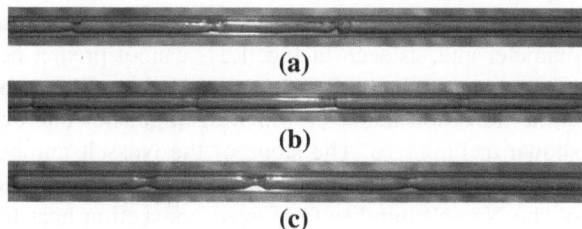

Fig. 1.21. Figures 1.24 and 1.25 show the observed flow patterns for these cases, liquid ring flow only. There was no observable change in the flow pattern over the range of gas-Reynolds numbers investigated. In Figs. 1.24 and 1.25, the turbulent wavy film seen in Figs. 1.22 and 1.23 is absent. The dominant effects of surface tension and liquid viscosity in the 222 and 140 μm channels prohibit turbulence and agitation of the gas–liquid interface, as suggested by Chung and Kawaji (2004), Kandlikar and Grande (2003), and Kawahara et al. (2002). The large asymmetric interfacial waves seen in the 506 and 334 μm channels become thin and axially symmetric in the 222 and 140 μm channels, due to the stronger influence of surface tension on the liquid film structure. That is to say, the separated air and water flow becomes laminar, and the mixing effect disappears, reducing heat transfer. To the best knowledge of the present authors, there is no quantitative information in the open literature on the relationship between heat transfer and the liquid film structure. Further research is required to determine this relationship.

Fig. 1.25 Two-phase flow
patterns within 145 μm
channel for $Re_L = 420$:
a $Re_G = 98$; **b** $Re_G = 144$;
c $Re_G = 321$ (Choo and Kim
2011)

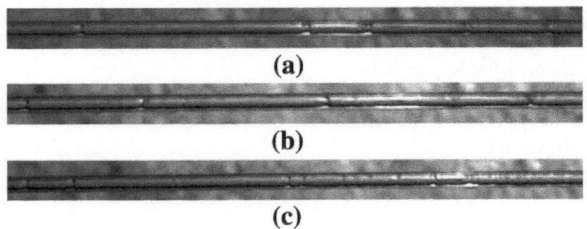

(a)

(b)

(c)

Equation 1.42 below is Chen's correlation (1966) for the two-phase forced convection Nusselt number:

$$\mathrm{Nu}_{TP} = 0.023\mathrm{Re}_L^{0.8}\mathrm{Pr}_L^{0.4}\left[2.35\left(0.213 + \frac{1}{X}\right)^{0.736}\right] \tag{1.47}$$

As evaluated by Ghiaasiaan (2008), Chen's correlation is perhaps one of the most widely acknowledged correlation for mini- and microchannels. It works well for water and has been applied to a variety of fluids. In particular, Chen's correlation is known to predict the forced convection contribution to two-phase heat transfer in macroscale channels. However, Chen's correlation does not include change in diameter and, as seen in Fig. 1.21, cannot predict heat transfer characteristics in microchannels that have channel diameters of less than 500 μm. In addition, there is the transition diameter which distinguishes microchannel from minichannel as shown in Fig. 1.26. The slope of the Nusselt number decreases with decreasing channel diameter as shown in Fig. 1.27. Although there are previous correlations of the Nusselt number for forced convection heat transfer (Steiner and Taborek 1992; Gungor and Winterton 1986; Li and Wu 1991; Kandlikar 1990), they cannot predict the heat transfer characteristics of microchannels. The correlations of Steiner and Taborek (1992), Gungor and Winterton (1986), and Liu and Wu (1991) give the Nusselt number as being independent of the channel diameter. Moreover, Kandlikar's correlation (1990) shows that the Nusselt number increases with decreasing the channel diameter. Recently, Choo and Kim (2011) developed a new correlation, as shown below, that accounts for the effect of diameter in two-phase microchannel flow.

$$\mathrm{Nu}_{TP} = 0.023\mathrm{Re}_L^{m}\mathrm{Pr}_L^{0.4}F \tag{1.48}$$

where the factor F is meant to represent $(\mathrm{Re}_{TP}/\mathrm{Re}_L)^m$, given by

$$F = C \cdot X^{-n} \tag{1.49}$$

$$m = 0.8 - 0.8\left[1 + e^{(d^* - 37)/7}\right]^{-1} \tag{1.50a}$$

$$C = 2.94 + 358 \cdot e^{-(0.1d^*)} \tag{1.50b}$$

$$n = 0.7 - 0.8\left[1 + e^{(d^* - 41)/2}\right]^{-1} \tag{1.50c}$$

Fig. 1.26 Variation in Nusselt number with channel diameter, showing the transition diameter (Choo and Kim 2011)

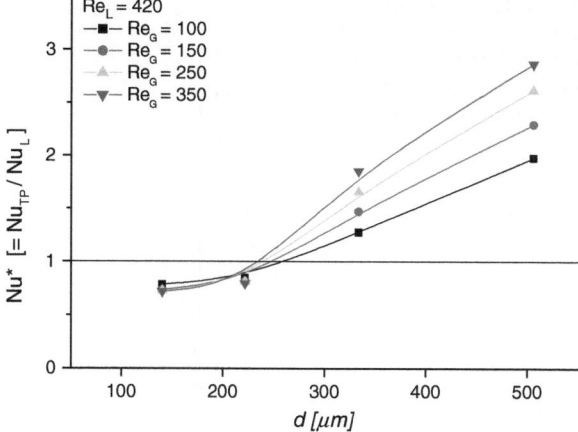

Fig. 1.27 Variation in Nusselt number with liquid flow rate at a fixed gas flow rate

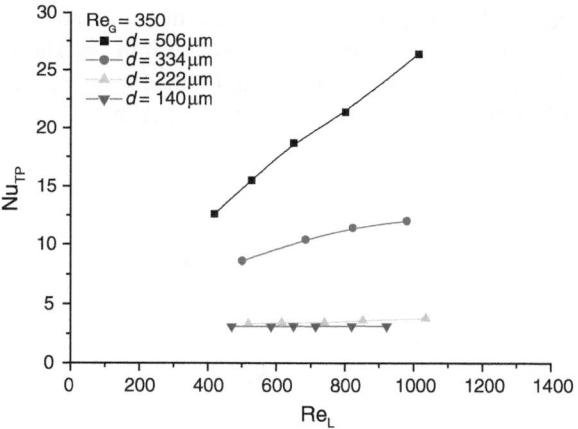

The dimensionless channel diameter is defined by $d^* = d/(\sqrt{\sigma/pg})$. It is worth noting that the functional form of Eq. (1.48) was chosen to match the form of Eq. (1.47). As shown in Eq. (1.50a), the exponent m of Eq. (1.48) goes to 0 as the dimensionless channel diameter decreases. As shown in Fig. 1.27, this means that the effect of liquid flow rate (Re_L) is diminished as the dimensionless channel diameter decreases. However, the exponent m of Eq. (1.48) goes to 0.8 as the dimensionless channel diameter increases. As shown in Eq. (1.49), the factor F has a form similar to the previous correlation, Eq. (1.47). As shown in Eq. (1.50c), the exponent on the Martinelli parameter n in Eq. (1.49) becomes negative as the dimensionless channel diameter decreases. As shown in Fig. 1.21, at smaller channel diameters this means that the Nusselt number decreases with increasing gas flow rate (Re_G). However, the exponent n of Eq. (1.50c) goes to 0.7 as the dimensionless channel diameter increases. Therefore, the present correlation,

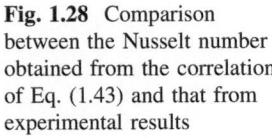

Fig. 1.28 Comparison
between the Nusselt number
obtained from the correlation
of Eq. (1.43) and that from
experimental results

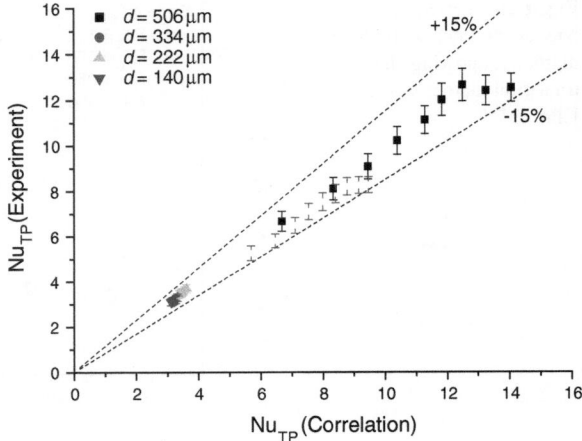

Eq. (1.48), becomes identical to the previous correlation, Eq. (1.47), as the
diameter becomes larger than 0.506 mm, as shown in Fig. 1.21. As shown in
Fig. 1.28, the correlation of Eq. (1.48) for microchannels of diameter greater than
140 μm is well matched with the experimental results within ±15 %.

Chapter 2
Force-Fed Microchannels for High Flux Cooling Applications

Abstract In this chapter, the working principles of force-fed microchannel (micro-grooved) heat exchangers (FFMHX) in single-phase and two-phase heat transfer modes are defined and the associated benefits of using these systems are highlighted. The literature on FFMHX is reviewed, and important conclusions are summarized. The main benefits of optimally designed next generation force-fed heat transfer (FFHT) configurations are the short, parallel microchannel network system and equal flow distribution among the channels through careful manifold design. The collective benefits include substantially higher available heat/mass transfer surface area, precise flow distribution among the channels, creation of thin film cooling in microchannels such that the thin film component dominates the flow regime, and low to moderate pressure drops while achieving record-high heat/mass transfer coefficients. The information presented in this chapter includes experimental evaluation of the thermal performance of FFMHXs in two-phase heat transfer mode using refrigerant R-245fa. Two distinct heat transfer trends were observed. At high hydraulic diameter, high mass flux, and high heat flux, the heat transfer coefficients had a slowly increasing trend with increase in heat flux and outlet quality. At low hydraulic diameters, low mass flux, and low heat flux, the heat transfer coefficients experience a bell-shaped behavior with a sharp increase at low vapor qualities until they reach the maximum peak point. Most recent research has focused on increasing the vapor quality at the exit of the evaporator for energy efficiency and optimum system performance. It was demonstrated that the tested FFMHX heat sink configuration can cool a heat flux of $q''_{base} = 1.23 \text{kW/cm}^2$ with a superheat of $\Delta T_{sat} = 56.2°C$ and pressure drop of only $\Delta P = 60.3 \text{ kPa}$. The results clearly suggest that under optimum design conditions, FFMHXs have the proven potential to achieve substantially higher heat transfer coefficients with low

M. Ohadi et al., *Next Generation Microchannel Heat Exchangers*,
SpringerBriefs in Thermal Engineering and Applied Science,
DOI: 10.1007/978-1-4614-0779-9_2, © The Author(s) 2013

to moderate pressure drops, and substantially less working fluid in circulation, and thus substantial compaction (weight/volume reductions), improved energy efficiency, and reduced capital and operational cost. When compared to the state-of-the-art thermal management/conventional systems.

2.1 Introduction

High heat flux cooling is required in many applications such as power electronics, plasma-facing components, high heat-load optical components, laser diode arrays, X-ray medical devices, and power electronics in hybrid vehicles. In general, the exposed area that needs to be cooled for these systems is limited, and the amount of heat that needs to be removed is extremely high, thus requiring cooling of high heat fluxes. While high heat flux cooling is essential for creating an efficient cooling system, there are usually also other system requirements, such as low thermal resistance, surface temperature uniformity, low pumping power, compact design, suitability for large area cooling, and compatibility for use with dielectric fluids. Currently, most of the advanced electronic components already generate heat fluxes exceeding 100 W/cm^2, while some future microprocessors and power-electronic components, such as high-power laser and electronic radar systems, have been projected to generate heat fluxes over 1,000 W/cm^2 (Mudawar 2001; Kandlikar 2005; Kandlikar and Bapat 2007). The increase in power density of the components has also created a need for advanced cooling technologies to achieve high heat dissipation rates in order to keep the electronic system at the desired working temperatures. At this point, traditional and well-known cooling methods will prove insufficient for such high heat fluxes.

The force-fed microchannel heat exchanger (FFMHX) concept (also known as the manifold microchannel heat sink) was first introduced by Harpole and Eninger (1991), and later studies reported that FFMHX can achieve heat transfer coefficients of 30–50 % (Kim and Chun 1998; Ryu and Choi 2003) higher than traditional microchannel heat sinks (TMHS) at the lowering pumping power. However, due to flow and their geometrically complex nature, FFMHXs in the past received less attention, and the flow and heat transfer associated with these heat sinks have not been studied in detail yet. This chapter provides an overview of the most recent progress in FFMHX and the basics of the heat transfer enhancement mechanism that makes this cooling technology unique. Differences from TMHS will also be highlighted.

Fig. 2.1 Schematic flow
representation of a typical
FFMHX (Cetegen 2010)

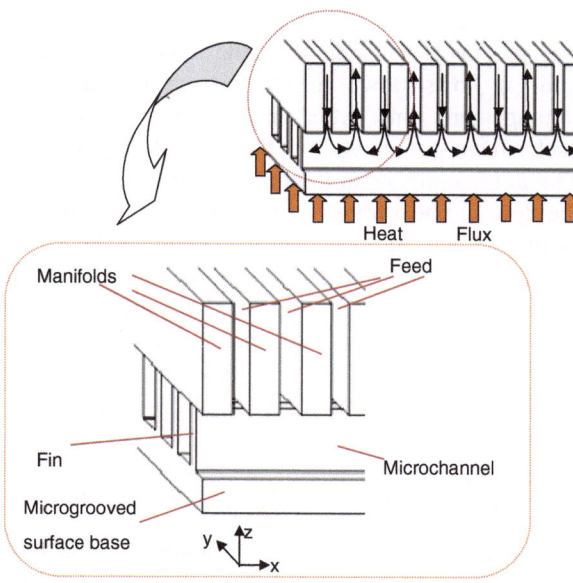

2.1.1 Working Principles

An FFMHX is basically a combination of a microgrooved surface and a system of
manifolds. The flow schematic of a typical FFMHX is shown in Fig. 2.1. The flat
side of the microgrooved surface is attached to the heat source, while the fins and
microchannels on the other side are in contact with the working fluid. A series of
parallel manifolds is located on the top of the microchannels and aligned perpendicularly to the fin structure. The manifolds do not contribute to heat transfer
significantly, their primary role being to distribute the fluid and provide structural
integrity to the FFMHX. Each gap between the two neighboring manifolds forms a
feed channel, which is used to direct the fluid in (inlet feed channel) or out (outlet
feed channel) of the microchannels. This gap can have similar or different
dimensions for inlet and outlet feed channels, depending on the desired design
configuration. From a design standpoint, compared with microchannels, the feed
channels have much greater hydraulic diameters and lower flow velocities. The
manifolds and the microgrooved surface are usually attached to each other by use
of compressive force that seals the tip of the fins to the bottom of the manifolds.

The inlet feed channels in the manifold structure are designed to deliver the
same flow resistance so that when they are connected to a common fluid source,
flow uniformity can be easily achieved. The flow in the feed channels can be
considered as flow between two parallel plates due to the usually high aspect ratio
of the feed channels. After entering and flowing along the feed channel, the fluid
encounters the microchannels and fins of the microgrooved surface. Here, the fluid
is forced to enter the microchannel from the top, perpendicularly to the microchannel direction. This entrance creates a flow area reduction and in turn increases

the flow velocity and decreases the static pressure at the entrance region of the microchannel. After entering the microchannel area, the flow starts to develop until it flows down to the bottom of the channel, eventually creating an impingement zone. After the stagnation point, the fluid splits into two streams, each stream turning 90° and flowing in the opposite directions in the microchannel. The fluid continues to flow a short distance in the microchannel and at this point the flow and heat transfer occur similarly to that in a typical TMHS. The distance of the straight microchannel is defined by the thickness of the manifolds. At the end of the straight microchannel, the fluid will make a second 90° turn, joining with the counterstream and leaving the area through the corresponding outlet feed channel. The exit from the microchannel to the outlet feed channel creates a pressure increase and velocity reduction due to an increase in flow area. This flow configuration is repetitive and results in the formation of arrays of short microchannels working in parallel.

The key geometrical arrangements and flow distributions that make FFMHX an effective heat transfer cooling system are as follows:

- The system pressure drop is decreased significantly due to the short flow length of the turning path in the microchannels. The system is, in fact, a network of short microchannels working in parallel, and therefore the total system pressure drop is the pressure drop of a single microchannel flow turn.
- FFMHX is suitable for cooling large areas and can be easily expandable in the x–y directions. By increasing the flow rate proportional to the base area expansion rate, the system pressure drop and effective heat transfer coefficient remain constant.
- FFMHX benefits from multiple inlet entrance effects. The area in the microchannel flow inlet region can yield very high heat transfer coefficients. This effect is the result of thermally developing flow in this region, which is associated with very thin boundary layers and low thermal resistance values. Multiple inlet regions enhance the overall heat transfer coefficient of the heat sink.
- Generally, electronic chips that need to be cooled are spaced closely together on the substrate, and there is limited space in the x–y direction for including additional equipment such as flow distributing manifolds. In this case, it could be more convenient to include additional parts in the z-direction, which makes FFMHX design favorable.

FFMHX design requires enhanced surfaces with alternating fins and channels, also known as microgrooved surfaces. These surfaces can be fabricated using several different methods, depending on the substrate material and geometric features. For example, the silicon microfabrication technique can create very fine surfaces with microchannels having hydraulic diameters on the order of microns. Micromachining is another process that can create microchannels using methods such as electron discharge machining (EDM) or laser ablation. These methods are the most suitable for metal substrates. Another fabrication method that is becoming more popular is micro deformation technology (MDT). This relatively new technology allows fabrication of enhanced surfaces with very high aspect

Fig. 2.2 Picture of a typical microgrooved surface profile fabricated with MDT (Thors and Zoubkov 2009)

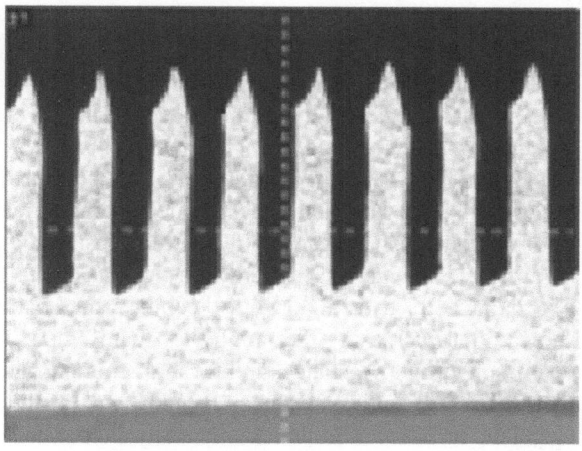

ratio channels and fins from a wide range of metals. MDT was initially developed to produce enhanced heat transfer tubes and was later adopted for flat surfaces. The fabrication principle of MDT is based on a continuous process of skiving and bending material on the top of a metal substrate. The major advantage of MDT is that it can be cost effective when adapted for mass production (Thors and Zoubkov 2009).

A cross-section of a typical microgrooved surface fabricated using MDT is shown in Fig. 2.2. The fin geometry created by MDT is usually slightly different from the fin structure of a traditional microgrooved surface fabricated usually by the silicon microfabrication technique. Based on the cutting tool geometry used during fabrication, MDT microgrooved surfaces have an unconventionally sharp fin-tip and a slightly bent fin geometry.

2.1.2 Survey of State of the Art Research in the Field

The earliest study to introduce and investigate heat transfer and fluid flow in an FFMHX is by Harpole and Eninger (1991). They considered microchannels and manifolds etched from silicon and an optional diamond face sheet attached between the heat sink and heat source for heat spreading and enhancing the surface temperature uniformity. By neglecting the entrance effects and convective heat transfer by assuming a low Reynolds number flow and constant Nusselt number, they formulated the flow and heat transfer using a 2D model. Based on this model, they performed a parametrical analysis of the microchannel geometry and flow by varying one parameter and keeping the other parameters constant. They concluded that when water is used as a coolant, heat fluxes on the order of 1,000 W/cm^2 can be cooled effectively with pressure drop values on the order of 1 bar. They also noted that although the pressure losses at the inlet and outlet to the microchannels were not considered in the model, they could be significant and needed to be included in a more detailed design.

Copeland (1995) analyzed flow and heat transfer in an FFMHX by assuming one-dimensional flow and building a simple model based on the hydrodynamically and thermally developing flow correlations in straight tubes. The heat sink material was assumed to be poorly conductive, and a uniform heat flux was applied to the base. Using the mathematical model, parametrical analyses were performed to investigate the variation of thermal resistance with flow and geometrical parameters. The analysis strategy was to investigate thermal resistance by varying two parameters at a time while keeping other parameters constant. The results showed that increasing the number of manifolds per unit length of heat sink can decrease thermal resistance, and for a given manifold number and pumping power there exists an optimum channel width and fin thickness. Copeland (1995) and Copeland et al. (1997) numerically analyzed the heat transfer and pressure drop in an FFMHX for 32 different cases with different geometries and inlet flow conditions. The computational domain was simplified by neglecting the conjugate heat transfer in the solid part of the microgrooved surface. Therefore, only the fluid flow was modeled, and constant heat flux or constant temperature boundary condition was applied to all heat transfer surfaces (microchannel base and fin surfaces). Their model also neglected the heat transfer from the tip of the fins and the pressure losses due to both flow contraction and expansion at the inlet and outlet to the microchannels.

Based on the numerical results, two different heat transfer characteristics were observed. The first heat transfer mode occurred at low channel velocities of 0.1 m/s. In these cases, the local heat transfer coefficients peaked at the inlet region and then gradually decreased to the end of the microchannel. The second mode of heat transfer was reported at high flow velocities of 1 m/s. Here, the local heat transfer coefficient was nonuniformly distributed along the channel walls, with the maximum close to the inlet region and two other local maxima at the channel bottom below the inlet and close to the outlet at the end of microchannel. Although not reported in their study, these different heat transfer characteristics are the result of flow impingement and secondary flow-induced vortices, which can create local maximum heat transfer coefficient regions. The study also compared the numerical results with 1D straight channel models, and it was found that such a simplified model failed to predict the pressure drop and thermal resistance values obtained by numerical simulations.

Kim and Chun (1998) experimentally investigated the thermal performance of FFMHXs for air cooling. Several experimental tests were performed for nine different copper microgrooved surfaces and two manifold systems with different geometrical dimensions. As a general trend for all cases, they found that thermal resistance is a strong function of flow rate and that thermal resistance decreases linearly with increasing pumping power on a log–log plot. Among the investigated geometrical parameters, microchannel width and manifold inlet and outlet feed channel widths were found to affect the thermal resistance the most. Interestingly, they also reported that for two heat sinks with identical microgrooved surfaces but different manifold designs, increasing manifold number per unit heat sink length increased thermal resistance at the same pumping power. This result conflicts with the work of Copeland (1995), where it was shown that increasing the manifold number had a beneficial effect on decreasing thermal resistance. The geometrical design that delivered the lowest thermal resistance was compared with a traditional microchannel heat sink, and it was found that the FFMHX design would work with 35 % less thermal resistance at the same pumping power.

Ng and Poh (1999) analyzed thermal performance of FFMHXs numerically for 16 cases with different geometrical dimensions and flow conditions. The numerical model was validated by the experimental results of Copeland (1995). They also formulated a 1D analytical model based on straight channel pressure drop and heat transfer correlations. The numerical model was compared with the analytical model for thermal resistance at similar geometrical and flow parametric input conditions. The comparison showed that the 1D analytical model could predict the thermal resistance values at an acceptable range only at very low Reynolds numbers ($Re < 50$). At higher Reynolds numbers, the analytical model largely under-predicted (up to 400 %) the thermal resistance values. Overall, it was concluded that simple 1D analytical models based on straight channel flow cannot predict the thermal performance of an FFMHX accurately.

Ryu and Choi (2003) analyzed the heat transfer and flow in an FFMHX by numerically solving the 3D Navier Stokes equations. Using a finite volume approach, their numerical model considered both the convective heat transfer in the channel and the heat transfer through the fins and base material. The base material selected was silicon, and water was used as the working fluid. By keeping the pumping power and manifold number constant and varying other parameters, they optimized the geometry using a steepest-descent technique to minimize the thermal resistance. They compared their results with TMHS and reported that the thermal resistance could be reduced by more than half, while the temperature uniformity on the heated wall could be improved by tenfold.

Jankowski and Everhart (2007) explored experimentally the possibility of reducing the thermal resistance of power electronics used in hybrid vehicles by integrating two different cooling systems involving microchannels. The first approach consisted of integrating an FFMHX to cool chips directly on a silicon substrate. The second approach used a TMHS with microchannels created in an aluminum nitride substrate. The thermal resistance values for both systems were measured at pressure drops of 17 and 35 kPa. The results show that the thermal resistance values for both systems were comparable at 17 kPa, and for 35 kPa FFMHX thermal resistance was lower by only about 12 % compared with the THMS design. The authors attributed this low difference to the nonoptimized geometries of FFMHX and suggested that the heat transfer geometry needed to be optimized for future potential improvement.

Xia and Liu (2008) investigated experimentally the effect of two different surfactants on pressure drop in a single-channel manifold microchannel with the goal of increasing the system drag reduction. They used sodium dodecyl sulphate and alkyl polyglycoside aqueous solutions as working fluids. They found that the measured drag reduction values were not significant for laminar flow regime, but that the pressure drop reduction was more effective in transition flow at high temperatures. The transition from laminar flow to transitional flow occured at a critical Reynolds number of 800.

Haller et al. (2009) investigated experimentally and numerically the pressure loss and heat transfer in bending and branching microchannels such as L-bends and T-joints for rectangular channels with a 90° turning angle. This geometrical configuration is similar to a unit cell of an FFMHX without the second turning bend at the end of the microchannel. Using water, they tested silicon-based channel flow for Reynolds numbers ranging from 10 to 3,000. Based on both numerical and experimental results, they reported that flow turning can enhance the heat transfer when fluid was redirected in the 90° bend. This effect was the result of flow vortices induced by secondary flows created by centrifugal forces. They also reported that the creation of such vortices also increased the pressure drop of the system due to higher viscous dissipation losses. The heat transfer efficiency and pressure drop characteristics were found to be always conflicting, with higher heat transfer performance always requiring higher pumping power. They also unsuccessfully attempted to model the pressure loss of the laminar flow in microchannel bends with vortices. They conceded that the main challenge of building such a model is the different flow characteristics observed at different flow conditions and channel geometries. For example, the creation of no vortices, or one pair or two pairs of vortices was strongly dependent on Reynolds number, shape, and aspect ratios of the channel.

Cetegen et al. (2007, 2008) performed experimental tests to determine the boiling heat transfer performance of FFMHX for different microgrooved surfaces and working fluids. They were able to show that force-fed cooling can cool up to 925 W/cm^2 with heat transfer coefficient of 130,000 W/m^{2}K using the nonaqueous refrigerant HFE-7100. These results demonstrated that FFMHX is a promising candidate for applications that require high heat flux cooling with high efficiency. Cetegen (2010) later proposed and tested a similar but improved test section by decreasing the measurement uncertainties and decreasing the two-phase flow instabilities. For two-phase FFMHX, three different heat sink designs incorporating microgrooved surfaces with microchannel widths between 21 and 60 μm were tested experimentally using refrigerant R-245fa, a dielectric fluid. It was demonstrated that FFMHX could cool a heat flux of 1.23 kW/cm^2 with a wall superheat of 56.2 °C and pressure drop of 60.3 kPa.

A summary of this literature survey is given in Table 2.1. Two major conclusions can be drawn from this survey. First, FFMHX has the potential to achieve higher thermal performance values than TMHS. However, a comparison between the two needs a more systematic approach. One substantive method would be to compare the thermal performance of FFMHX with other conventional cooling technologies at optimum design and working conditions. Second, since all the parametric studies and optimization techniques used in previous studies are based on single-objective optimization methods, in order to obtain the real optimum

Table 2.1 Summary of related work on single-phase FFMHXs

Authors	Type of work	Microgrooved surface material (TSM) and working fluid (WF)	Microchannel hydraulic diameter (micron)	Reynolds number and flow regime	Parametrical study or optimization performed	Conclusion
Harpole and Eninger (1991)	2D analytical model	TSM: silicon WF: water	15–66	15–400 (laminar flow)	Yes	FFMHX could be used effectively for cooling high fluxes
Copeland (1995)	1D analytical model	TSM: low conductive material WF: water	8–64	(laminar flow)	Yes	There exist optimum points for microgrooved surface and manifold dimensions
Copeland and Behnia (1997)	3D numerical simulation using fluent	TSM: silicon WF: water	113–226	18–485 (laminar flow)	Yes	1D models based on straight channel flow cannot predict thermal performance of FFMHX
Kim and Chun (1998)	Experimental	TSM: copper WF: air	700–2000	(laminar flow)	Yes	FFMHXs can perform 30 % better compared with TMHS thermal performance
Ng and Poh (1999)	3D numerical simulation using ansys/flotran	TSM: silicon WF: water	113–226	10–800 (laminar flow)	Yes	1D models based on straight channel flow can under-predict thermal performance of FFMHX by 400 %
Ryu and Choi (2003)	3D numerical model	TSM: silicon WF: water	10–60	1–100 (laminar flow)	Yes	Compared with TMHS thermal performance, FFMHX can perform 50 % better with ten times better surface temperature uniformity
Jankowski and Everhart (2007)	Experimental	TSM: silicon and AlN WF: water	40–760	(laminar flow)	No	FFMHX needs to be optimized to achieve comparable better performance

(continued)

Table 2.1 (continued)

Authors	Type of work	Microgrooved surface material (TSM) and working fluid (WF)	Microchannel hydraulic diameter (micron)	Reynolds number and flow regime	Parametrical study or optimization performed	Conclusion
Xia and Liu (2008)	Experimental	TSM: silicon WF: aqueous surfactant solution	200	100–3500 (laminar and transitional flow)	No	Surfactants can decrease pressure drop at high temperatures and Reynolds numbers
Haller and Woias (2009)	Experimental and 3D numerical model	TSM: silicon WF: water	300–1200	10–3000 (laminar and transitional flow)	No	Creation of secondary flows at high Re numbers can enhance heat transfer

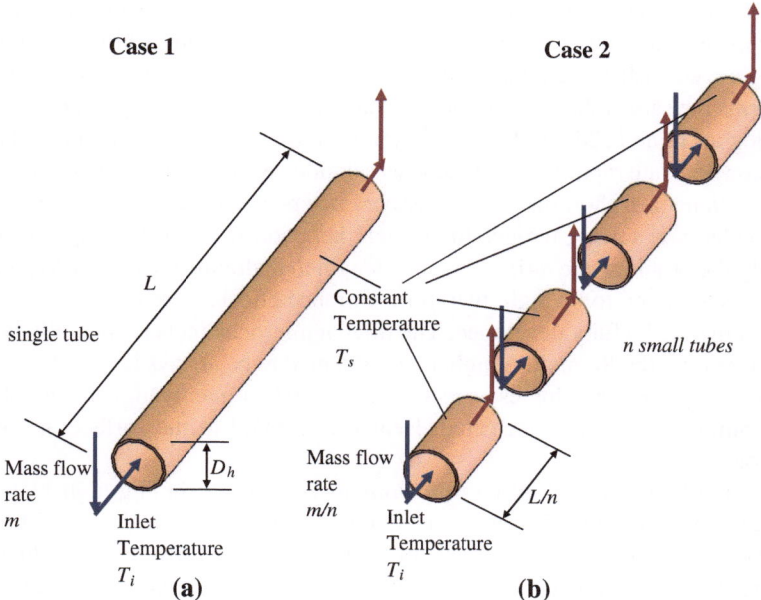

Fig. 2.3 Schematic of flow in **a** a single long channel and **b** multiple short channels (Cetegen 2010)

designs, a multiobjective optimization technique is necessary. With such a technique, the thermal performance could be optimized based on the most important objective functions, which in this case are heat transfer coefficient and pumping power. The objective should be to maximize the heat transfer coefficients while minimizing the pumping power.

2.2 Heat Transfer Analysis of FFMHX

Simple analyses can demonstrate the thermal performance advantages of FFMHX over traditional microchannel single-phase flow. Using heat transfer and pressure drop formulations for internal flow, the reduction in pumping power of the force-fed heat transfer (FFHT) flow configuration can be estimated. The procedure and assumptions used in this analysis are discussed in the following sections.

2.2.1 Heat Transfer and Pumping Power in Short Channels

Consider the two flow configurations shown in Figs. 2.3a, b. Case 1, shown in Fig. 2.3a, depicts flow in a single long tube with hydraulic diameter of D_h, tube length L, mass flow rate $\dot{m}$, constant surface temperature T_s, and inlet temperature T_i. This case is analogous to a single channel in a TMHS with a highly conductive microgrooved surface base material. The second case, Case 2, shown in Fig. 2.3b, represents the same tube divided into n equal length, short tube segments. This case is analogous to an FFMHX flow configuration formed from a single microchannel. The mass flow rate running in each tube was divided equally, resulting in a mass flow rate of $\dot{m}/n$ in each short tube. All other parameters, such as hydraulic diameter and inlet, outlet, and surface temperatures, were kept constant. At this stage, to further simplify the problem, the ratio of channel length to hydraulic diameter L/D_h for both cases was assumed to be high; therefore, entrance effects were neglected and flow was assumed to be fully developed. The flow in microchannels is generally laminar flow; therefore, the Reynolds number was assumed to be always below $Re < 2{,}300$. Also, the major pressure losses were assumed to be dominantly higher, and inlet and outlet minor losses were neglected. Heat was applied to the surface at constant temperature T_s.

It should be noted that in the following analysis, by assuming high L/D_h ratio, the developing regions in the short microchannels were neglected. This is a conservative approach since short microchannels in FFMHX can actually be optimized to utilize this heat transfer augmentation method, and efficiency can be improved dramatically. But the objective of the analysis in this chapter is to show the distinctive advantages of short channel systems over the traditional long channel ones, which will be proven even without including the developing region. With the addition of the developing heat transfer region, the discrepancy between the two systems would be even more pronounced.

Usually, the most important performance parameters considered in a heat sink design are overall heat transfer coefficient between the heated surface and coolant temperatures, and the pumping power. While the overall heat transfer coefficient is desired to be as high as possible to increase heat transfer efficiency, the pumping power values need to be minimized. Since these two parameters conflict, a reasonable comparison can be made by keeping one parameter constant and performing the comparison based on the other. Here, the overall heat transfer coefficients for Case 1 and Case 2 will be kept constant, and the pumping power will be evaluated.

It can be shown that with constant surface temperature assumption, the outlet fluid temperature $T_{o,1}$ for Case 1 can be calculated from (Incropera and DeWitt 2002) as:

$$\frac{T_s - T_{o,1}}{T_s - T_i} = \exp\left(-\frac{\pi D_h L}{\dot{m}c_p}\bar{h}\right) \tag{2.1}$$

where $\bar{h}$ is the average heat transfer coefficient. For the current case, this value is constant along the tube and is equal to:

$$\bar{h} = h_z = Nu_T\frac{k}{D_h} = 3.66\frac{k}{D_h} \tag{2.2}$$

Here $\mathrm{Nu}_T = 3.66$ is the Nusselt number defined for fully developed flow in a tube with constant wall temperature (Incropera and DeWitt 2002), and k denotes the thermal conductivity of the working fluid. The heat transfer to the fluid is then calculated based on logarithmic mean temperature difference:

$$q_{\mathrm{Case1}} = \bar{h}(\pi D_h L) \frac{(T_s - T_{o,1}) - (T_s - T_i)}{\ln \frac{T_s - T_{o,1}}{T_s - T_i}} \tag{2.3}$$

Similarly, for each short tube defined in Case 2 the outlet fluid temperature $T_{o,2}$ can be calculated as

$$\frac{T_s - T_{o,2}}{T_s - T_i} = \exp\left(-\frac{\pi D_h L/n}{\dot{m}/n \, c_p} \bar{h}\right) = \exp\left(-\frac{\pi D_h L}{\dot{m} c_p} \bar{h}\right) \tag{2.4}$$

The right-hand sides of Eqs. (2.1) and (2.4) are identical, which indicates that for the two cases the outlet temperatures are equal. The total heat transfer in Case 2 is then evaluated by substituting $T_{o,1} = T_{o,2}$ and summing the heat transfer surface as:

$$
\begin{aligned}
q_{\mathrm{Case2}} &= \sum_{i=1}^{n} \bar{h}(\pi D_h L/n) \frac{(T_s - T_{o,2}) - (T_s - T_i)}{\ln \frac{T_s - T_{o,2}}{T_s - T_i}} \\
&= \bar{h}(\pi D_h L) \frac{(T_s - T_{o,1}) - (T_s - T_i)}{\ln \frac{T_s - T_{o,1}}{T_s - T_i}}
\end{aligned}
\tag{2.5}
$$

Equations (2.3) and (2.5) are equal, which indicates that the heat transferred from both the single-tube configuration of Case 1 and the multiple tubes with the same hydraulic diameter and heat transfer area of Case 2 is the same. Since inlet temperature is constant and equal for both cases, the heat sink heat transfer coefficients will be the same as well.

The pumping power for a control volume is the product of volumetric flow rate and pressure drop:

$$P_{\mathrm{pump}} = \Delta P \cdot \frac{\dot{m}}{\rho} \tag{2.6}$$

The major pressure losses for laminar flow in a tube can be calculated as:

$$\Delta P = f \frac{L}{D_h} \rho \frac{V^2}{2} \tag{2.7}$$

where $V = \frac{4\dot{m}}{\rho \pi D_h^2}$ is the mean fluid velocity in the tube and the pressure loss coefficient for fully developed laminar flow is defined as:

$$f = \frac{64}{Re} = \frac{64 \, \mu}{\rho V D_h} \tag{2.8}$$

Substituting Eqs. (2.7) and (2.8) into Eq. (2.6), the total pumping power required for Case 1 and Case 2 can be calculated as:

$$P_{\text{pump},1} = \left(\frac{128}{\pi} \frac{1}{D_h^4} \frac{\mu}{\rho^2}\right) L \dot{m}^2 \tag{2.9}$$

$$P_{\text{pump},2} = \sum_{i=1}^{n} \left(\frac{128}{\pi} \frac{1}{D_h^4} \frac{\mu}{\rho^2}\right) L/n(m/n)^2 \tag{2.10}$$

Based on these two equations the pumping power reduction can be calculated as:

$$\frac{P_{\text{pump},2}}{P_{\text{pump},1}} = \frac{1}{n^2} \tag{2.11}$$

Equation (2.11) demonstrates that for the same overall heat transfer coefficient, the split-flow configuration with multiple inlet and outlets has the potential to decrease the pumping power proportional to the square of the number of divisions. In other words, as long as the assumptions made above are valid, increasing the number of small channels by manifolding the flow always has a positive effect on decreasing the power consumption of the cooling system.

2.2.2 Heat Transfer and Pressure Drop for Thermally and Hydrodynamically Developing Flow

For the analysis performed in the previous section, it was assumed that the flow was fully developed along the whole tube. This assumption is not valid when the tube length over the tube hydraulic diameter ratio (L/D_h) is small. In this case, the heat transfer and momentum transfer occur mostly in the entrance region of the channel, where the hydrodynamic and thermal boundary layers are developing. Therefore, the heat transfer coefficient and friction factor in the entrance region depend on the distance from the entrance, Reynolds number, and fluid physical properties. These parameters may have a significant impact on thermal performance of FFMHX, where the heat transfer occurs mostly in the entrance region and the flow is hydrodynamically and thermally developing.

Assuming uniform velocity profile at the inlet, the hydrodynamic entry length $L_{\text{fd},h}$ for laminar flow in a tube is obtained from the following relation given by Langhaar (1942):

$$\left(\frac{L_{\text{fd},h}}{D}\right)_{\text{lam}} \approx 0.05 \, Re. \tag{2.12}$$

The friction factors in the hydrodynamically developing region are higher than those defined for fully developing flow. The skin friction and the additional momentum rate change due to change in the velocity profile are added together to

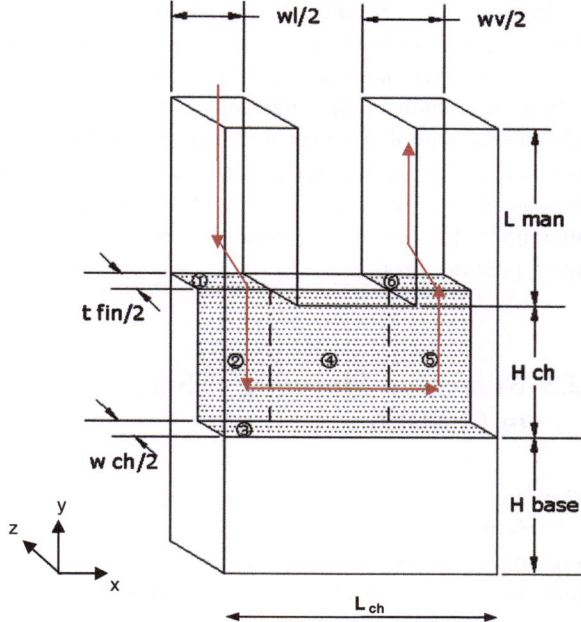

Fig. 2.4 Numerical computational domain (Cetegen 2010)

define the apparent friction coefficient. For circular tubes, Shah and London (1978) proposed the following correlation:

$$f_{app}Re = 344(x^+)^{-0.5} + \frac{\frac{1.25}{x^+} + (fRe)_{fd} - 344(x^+)^{-0.5}}{1 + 2.12 \times 10^{-4}(x^+)^{-0.2}} \quad (2.13)$$

where $(fRe)_{fd} = 64$ is defined for fully developed flow and x^+ is the nondimensional channel length defined as:

$$x^+ = \frac{L}{ReD} \quad (2.14)$$

For thermally developing laminar flows, Kays and Crawford (1980) proposed the following equation to calculate the thermal entry length $L_{fd,t}$:

$$\left(\frac{L_{fd,t}}{D}\right)_{lam} \approx 0.05 Re\,Pr \quad (2.15)$$

For the combined hydrodynamically developing region and thermal entry region for laminar flow in a circular tube, the average Nusselt number is given by Sieder and Tate (1936):

$$\overline{Nu} = 1.86 \left(\frac{Re\,Pr}{L/D}\right)^{1/3} \left(\frac{\mu}{\mu_s}\right)^{0.14} \quad (2.16)$$

Equations (2.13) and (2.16) demonstrate that in the entry region, both heat transfer coefficients and pressure drop increase compared to fully developed flow. This result may have an important impact on thermal performance of FFMHX, where the length of the channels is usually comparable to hydraulic diameters. Given that pumping power and heat transfer coefficients are the most important heat sink design parameters, both hydrodynamic and thermal entry effects need to be considered. For water and liquid working fluids with Prandtl numbers greater than unity, the benefit of increased heat transfer is usually higher than the penalty from pressure drop.

2.3 Numerical Modeling of Single-Phase Heat Transfer in FFMHX

Building an exact analytical heat transfer model of a force-fed microchannel heat sink is not practical due to the geometry and flow complexities mentioned in previous sections. The most efficient and feasible way to analyze the heat transfer and flow in such heat sinks is to perform numerical simulations. To decrease computational time, however, several assumptions need to be made. For example, the numerical modeling of the real-scale complete heat sink shown in Fig. 2.1 using CFD tools is neither feasible nor practical and presents real challenges in terms of computational time. A practical solution for this problem is to define a computational domain consisting of a much smaller but repetitive part of the real-scale heat sink. A sample computational domain is shown in detail in Fig. 2.4. Due to the repetitive nature of the computational domain over the entire heat sink on the y–z and x–y planes, symmetric boundary conditions can be used at the boundary surfaces located at these planes. The model is a combination of the microgrooved surface modeled as solid material and the fluid that flows through the feed channels and the microchannels. Because the microgrooved surface is symmetrical, it is modeled as the base and half of a fin. The microchannel is modeled as half of a channel, and similarly the inlet and outlet feed channels are considered only half of the channel width.

The computational domain shown in Fig. 2.4 includes several assumptions:

- The flow rate in each feed channel is steady and equal. The inlet feed channels of an FFMHX have identical geometries, and they are fed from the same pressure source. The outlet feed channels are formed by the gap between two neighboring manifolds, and therefore the pressure loss for each unit cell is constant. The only exception is the feed channels at the edges (the first and last feed channels), where there is no counter-manifold to create symmetry.
- When using microgrooved surfaces fabricated by MDT, the fins and channels of the microgrooved surfaces are usually not straight and may have a slightly bent geometry, as shown in Fig. 2.4, and the fins may end with an unconventionally

Table 2.2 Boundary conditions applied to computational domain

Location of applied boundary condition	Boundary condition
$x = 0$	Symmetry
$x = L$	Symmetry
$z = 0$	Symmetry
$z = t_{\text{fin}}/2 + w_{\text{ch}}/2$	Symmetry
$y = 0$	Uniform heat flux of $q'' = 1\ \text{kW/cm}^2$
$y = H_{\text{base}} + H_{\text{ch}} + L_{\text{man}}$	Specified inlet mass flow rate
$0 < x{<}w_i/2$	
$0 < z < t_{\text{fin}}/2 + w_{\text{ch}}/2$	
$y = H_{\text{base}} + H_{\text{ch}} + L_{\text{man}}$	Pressure outlet (zero outlet static pressure)
$(L - w_v/2) < x{<}L$	
$0 < z < t_{\text{fin}}/2 + w_{\text{ch}}/2$	

sharp fin tip. Here, the microgrooved surface geometry can be simplified by assuming straight fin geometry with flat fin tips.

- The heat flux applied from the bottom of the microgrooved surface should be considered constant. Simulating a nonuniform heat flux distribution is not feasible, since any variation in the X and Z directions will violate the symmetry conditions.
- The thermal properties (thermal conductivity) of the metal substrate are isotropic. Similarly, the symmetry boundary condition can be applied only for solids with isotropic thermal properties in the X and Z directions.
- The heat transfer through the manifolds is neglected, and the manifolds are considered adiabatic. There are two means of possible heat transfer in the manifolds. The first mechanism is the heat transfer through the incoming and outgoing fluid streams in the neighboring inlet and outlet feed channels. Here, a temperature gradient forms due to the temperature difference in the fluid streams, where fluid leaving the heat sink is hotter than the inlet steam due to the energy gained during heat transfer on the microgrooved surface. The heat transfer caused by this temperature gradient can be neglected by assuming that manifolds are made of poor conductive material such as low conductivity plastic. The second possible heat transfer mechanism is the thermal conduction through the tip of the fins to the manifold. In many practical applications, the manifolds are not bonded to the microgrooved surface, but rather kept in place by compressive forces. This configuration allows the microgrooved surfaces to be cleaned by easily disassembling the heat sinks in case of fouling and microchannel clogging. As shown in Fig. 2.4, the fin tips usually have a sharp edge, which creates a line of contact when compressed with the manifold top face, thereby creating a relatively high thermal contact resistance. To further simplify the problem and eliminate uncertainties associated with linear thermal contact resistance, an adiabatic manifold should be assumed for the numerical simulations.

2.3.1 Numerical Simulation of a Sample FFMHX

As an example, a typical FFMHX was numerically simulated and analyzed. The purpose of these numerical simulations was to demonstrate the working principle and to provide an initial design concept. The selected sample FFMHX configuration consisted of a microgrooved surface with channel height of $H_{ch} = 480$ μm, fin thickness of $t_{fin} = 48$ μm, microchannel width of $w_{ch} = 72$ μm, and base thickness of $H_{base} = 400$ μm, and manifold system with inlet feed channel width of $w_l = 400$ μm, outlet feed channel width of $w_v = 400$ μm, and total channel length of $L_{man} = 2$ mm. The microgrooved surface material and working fluid were copper and water, respectively.

2.3.1.1 Boundary Conditions and Numerical Domain

A complete list of applied boundary conditions is given in Table 2.2, based on the coordinate system shown in Fig. 2.4. Boundaries not included in the table were selected as adiabatic wall as default. The specified boundary condition for the inlet mass flow rate uses flow rate as input and calculates the static pressure based on the flow field. Therefore, the inlet static pressure is not known as *a priori* until the convergence is obtained. Similarly, the outlet boundary condition is specified as static pressure, and the total pressure is calculated after convergence is obtained.

The analyses were based on two output functions, the heat transfer coefficient and pumping power. Once the numerical convergence was achieved, these functions were evaluated based on post-processing data obtained from the flow field. The overall heat transfer coefficient was defined for the temperature difference between average surface temperature and inlet fluid temperature:

$$h = \frac{q''}{\overline{T}_{surf} - \overline{T}_{in}} \tag{2.17}$$

where the temperature values were area averaged over the surface as:

$$\overline{T} = \frac{1}{A} \int T \cdot dA \tag{2.18}$$

The pumping power was calculated for the unit base area. Pumping power has been defined by many researchers (Choo and Kim 2010b, c; Choo et al. 2012) as the product of total pressure difference between inlet and outlet boundaries and volumetric flow rate passing through the unit base area heat sink:

$$P'_{pump} = \frac{\dot{m}(\overline{P_o} - \overline{P_i})}{\rho\left[L\left(\frac{t_{fin}}{2} + \frac{w_{ch}}{2}\right)\right]} \tag{2.19}$$

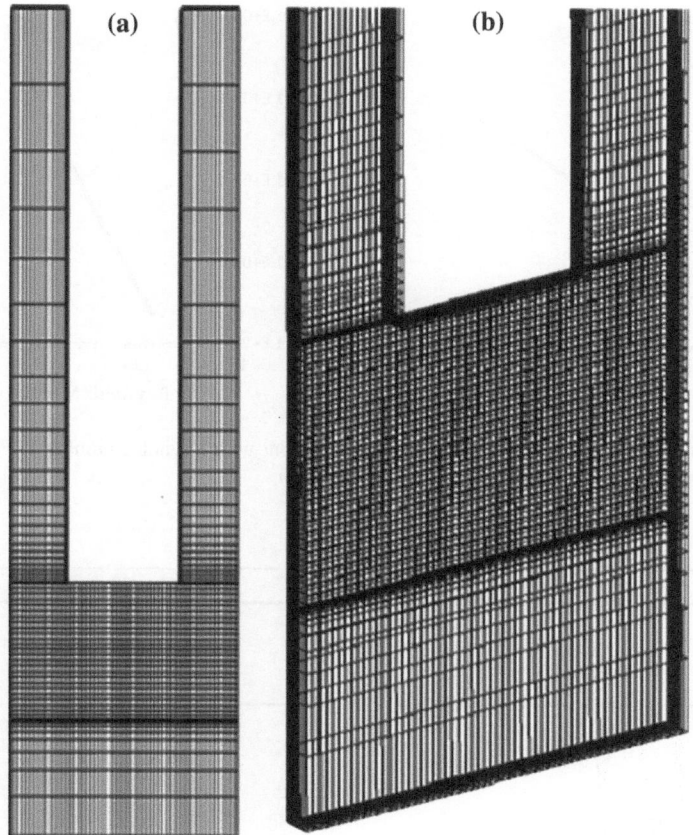

Fig. 2.5 a Front view and **b** perspective view for grid generated for $H_{ch} = 480$ µm, $w_{ch}/2 = 36$ µm, $t_{fin}/2 = 24$ µm, $L = 800$ µm, $w_l/2 = 200$ µm, $w_v/2 = 200$ µm, $L_{man} = 2$ mm, $H_{base} = 400$ µm (Cetegen 2010)

where the pressure values are also area averaged at inlet and outlet:

$$\overline{P} = \frac{1}{A} \int P \cdot dA \qquad (2.20)$$

The pumping power definition in Eq. (2.19) is practical for the extrapolation of pumping power for heat sinks with different base areas. In this case, for constant microchannel mass flux and microgrooved surface geometry, the pressure difference for the heat sink will remain the same, while the flow rate will increase linearly by increasing the heat sink base area.

Since heat transfer and pumping power are functions of flow rate, the flow conditions need to be defined. This is accomplished by introducing the Reynolds number based on flow in the straight channel part of the microchannel defined as:

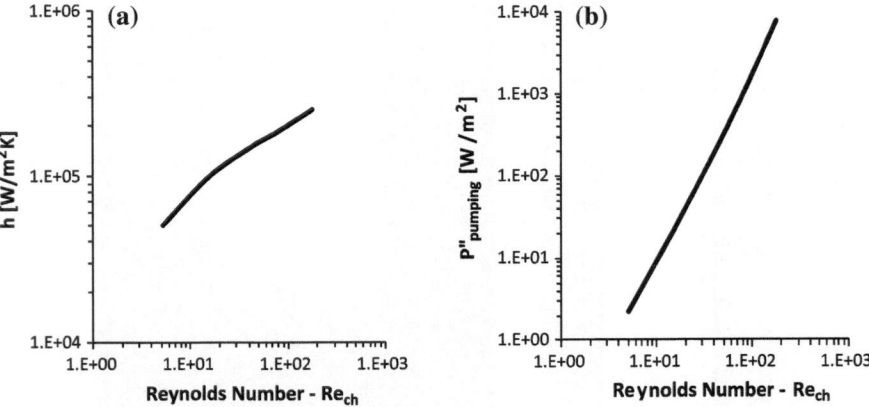

Fig. 2.6 a Variation of effective heat transfer coefficient with Reynolds number; **b** variation of pumping power with Reynolds number (Cetegen 2010)

Table 2.3 Numerical simulation results

	Unit	Case #1	Case #2	Case #3
Re	(–)	5	43	177
h	(W/m²K)	50,022	150,035	25,0094
P''_{pump}	(W/m²)	2.2	227.3	7888.8

$$Re = \frac{V_{ch}D_h}{v} \qquad (2.21)$$

where

$$V_{ch} = \frac{\dot{m}}{\rho H_{ch}w_{ch}} \qquad (2.22)$$

and

$$D_h = \frac{4H_{ch}w_{ch}}{2(H_{ch} + w_{ch})} \qquad (2.23)$$

Here, $\dot{m}$ is the total mass flow rate, V_{ch} is the average velocity and D_h is the hydraulic diameter, all defined for the flow in the straight part of the microchannel. All other geometrical dimensions are shown in Fig. 2.4.

The resulting computational grid is shown in Fig. 2.5. A grid independency study was performed by selecting a base case and refining mesh to generate five different grids, each consisting of 35,840, 48,576, 68,850, 98,604, and 137,280 cells. The heat transfer coefficient and pumping power per unit heat sink area were used as parameters for checking the grid independency. The results indicate that the case with 68,850 cells is reasonably accurate for the current study. When the

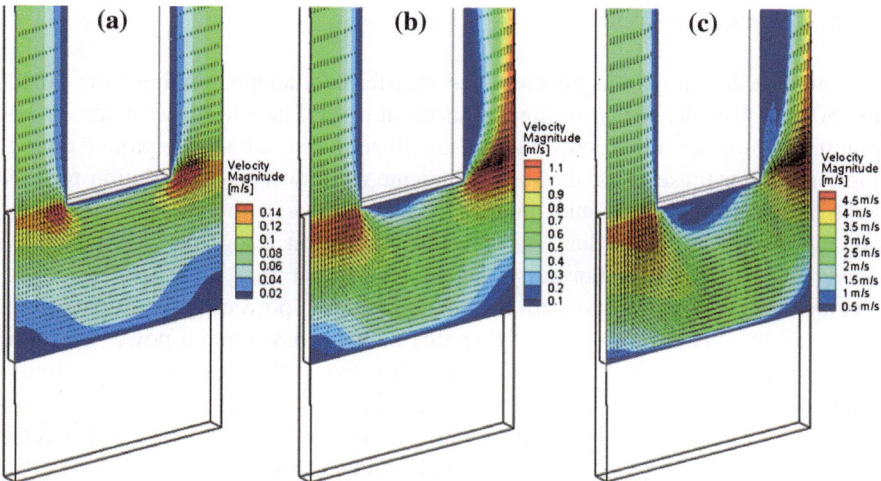

Fig. 2.7 Velocity vectors and velocity magnitude distribution at $z = 0$ for **a** Case #1 at $h = 50,000$ W/m^{2}K; **b** Case #2 at $h = 150,000$ W/m^{2}K; **c** Case #3 at $h = 250,000$ W/m^{2}K (Cetegen 2010)

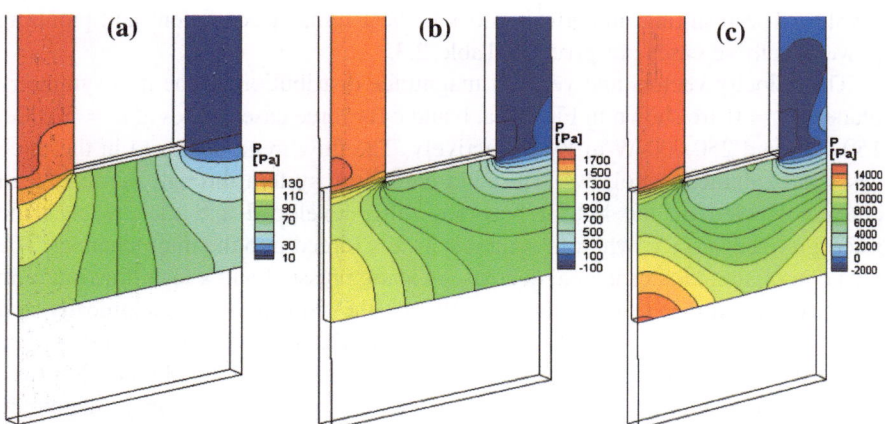

Fig. 2.8 Static pressure distribution at $z = 0$ for **a** Case #1 at $h = 50,000$ W/m^{2}K; **b** Case #2 at $h = 150,000$ W/m^{2}K; **c** Case #3 at $h = 250,000$ W/m^{2}K (Cetegen 2010)

number of cells was increased by almost two times, the variations in heat transfer coefficient and pumping power were calculated to be 0.29 and 0.47 %, respectively.

For all calculated cases, the convergence factor for continuity, three momentum equations, and energy equations were set as 1e-5, 1e-5, and 1e-11. The computational run time was 25 min on a PC with an Intel Pentium D 3.4 Ghz processor and 2 GB of memory.

2.3.1.2 Numerical Results

The numerical simulations covered heat transfer coefficients ranging from 50,000 to 250,000 W/m^{2}K. The resulting variations of overall heat transfer coefficient and pumping power with Reynolds number are given in Fig. 2.6a and b plotted on log–log charts. For these plots, the results indicate an almost linear variation for both heat transfer coefficients and pumping power values with increased flow rate. Having a slope different than one, the linear trend on a log–log plot indicates a power-law dependence. For fully developed laminar flow in a straight micro-channel with constant cross-section area, the pumping power is proportional to the square of the channel velocity or Reynolds number, indicating a power-law trend with power coefficient of two. On the other hand, the pumping power curve obtained in Fig. 2.6b has a power-law coefficient of ≈ 2.4. This in turn may suggest the influence of strong developing flow, secondary flows or other recirculation zones not present in straight microchannel flow.

To analyze the possible effects of these parameters, three different specific cases resulting in heat transfer coefficients of 50,000, 150,000, and 250,000 W/m^{2}K were selected as main case studies. The three cases are labeled Case #1, Case #2, and Case #3 and represent the low, medium, and high heat transfer coefficient designs. For each case, the mass flow rate was adjusted to give the targeted h value. The resulting numerical values of heat transfer coefficient and pumping power for these cases are given in Table 2.3.

The velocity vectors and velocity magnitude distribution on the mid-symmetry plane of $z = 0$ are shown in Fig. 2.7a, b and c for three case studies at $h = 50,000$, 150,000, and 250,000 W/m^{2}K, respectively. The flow trend observed in the three cases is explained as follows. The fluid enters the system through the inlet feed channel located on the left. It becomes fully developed and then reaches the microgrooved surface, where it contracts as it is forced into the microchannel. The area contraction creates a local decrease in static pressure and a local maximum in velocity magnitude. Due to entrance effects, the flow is hydrodynamically and thermally developing, starting with the inlet into the microchannel. The flow then turns 90° to the right as it undergoes an area expansion due to the higher flow area of the microchannel compared to the inlet flow area. The velocity magnitude distribution in this straight part of the microchannel is stratified in the y direction and with little variation in the x direction. The fluid exits the microchannel by making a second 90° turn and expanding into the outlet feed channel. A second velocity maximum is present just before the exit from the microchannel due to the flow area contraction–expansion effect.

While the results of the three cases shown in Fig. 2.7 have several common flow characteristics as discussed above, they differ in the physical mechanisms that characterize the flow. For example, at low Reynolds number flows (Fig. 2.7a), the inertial forces of the fluid entering the microchannel are comparable to the viscous forces in the microchannel. Therefore, the inertial forces of the incoming fluid are insufficient to push the fluid to the bottom of the microchannel and to create an impingement zone. Instead, due to the significant effect of viscous forces, the fluid

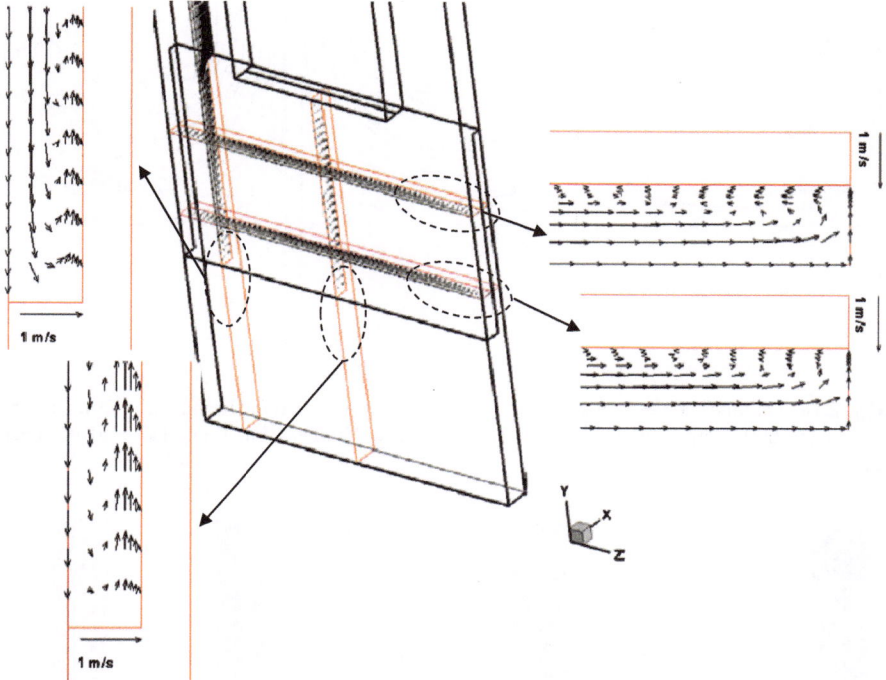

Fig. 2.9 Velocity vectors created by secondary flows at several cross-sections in the computational domain for Case #3 at $h = 250,000$ W/m²K and De $= 61$ (Cetegen 2010)

follows the path with the least friction losses, which creates a bypass zone just under the manifold between the inlet and outlet feed channels. The bypass zone in turn creates stratification in velocity distribution in the y direction with high velocities accumulated on the top and low velocity flow close to the bottom of the microchannel. The flow maldistribution is even more pronounced in the bottom microchannel regions under the inlet and outlet manifolds with the creation of "dead zones" with velocities below 0.02 m/s. This effect can be also explained by studying the static pressure distribution along the microchannel, as given in Fig. 2.8a for low Reynolds number flows. The static pressure distribution is uniform, and starting from the microchannel inlet, it decreases gradually along the microchannel length until it reaches the outlet zone. Due to the fluid bypass effect present at the top region, the fluid is not impinging the channel bottom, and no pressure increase is observed in this region.

Increasing the Reynolds number, by increasing the flow rate (Figs. 2.7b, c), leads to a different flow trend. First of all, the inertial force of the fluid entering the microchannel is high enough to push the liquid to the bottom of the microchannel and to create an impingement zone. The fluid first decelerates as it approaches this stagnation zone, turns the 90° bend, and then starts to accelerate again as it moves toward the straight part of the microchannel. The static pressure in the

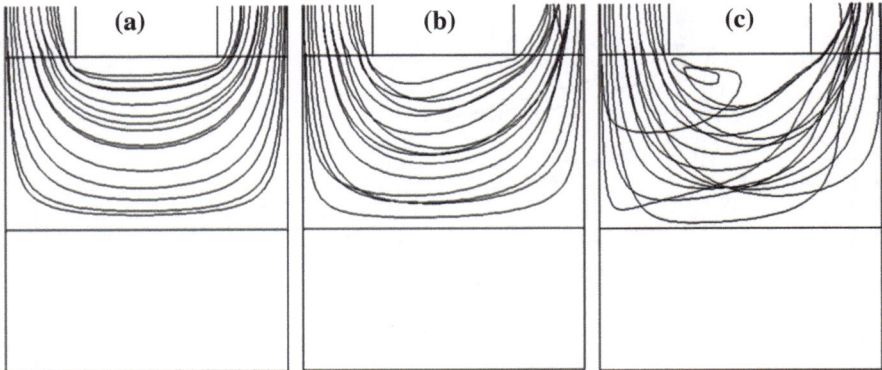

Fig. 2.10 Pathlines at $z = 0$ plane for **a** Case #1 at $h = 50,000$ W/m²K and De $= 2$; **b** Case #2 at $h = 150,000$ W/m²K and De $= 16$; **c** Case #3 at $h = 250,000$ W/m²K and De $= 61$ (Cetegen 2010)

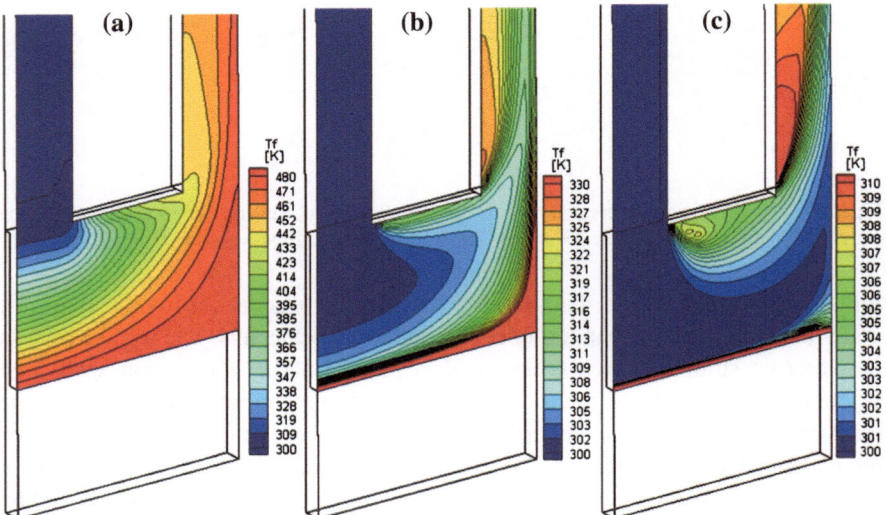

Fig. 2.11 Working fluid temperature distribution at $z = 0$ for **a** Case #1 at $h = 50,000$ W/m²K; **b** Case #2 at $h = 150,000$ W/m²K; **c** Case #3 at $h = 250,000$ W/m²K (Cetegen 2010)

impingement zone increases due to momentum exchange of this turning fluid (Figs. 2.8b, c). Second, the high inlet velocities and sharp corners of the manifold lead to flow separation that creates a recirculation zone under the manifold. The recirculation zone grows by increasing the Reynolds number. Interestingly, the velocity stratification for this case is opposite the low Reynolds number case. There is a high velocity core close to the bottom of the microchannel and low velocity recirculation zone at the top of the microchannel. A second recirculation zone is present at the bottom of the microchannel zone, under the outlet feed

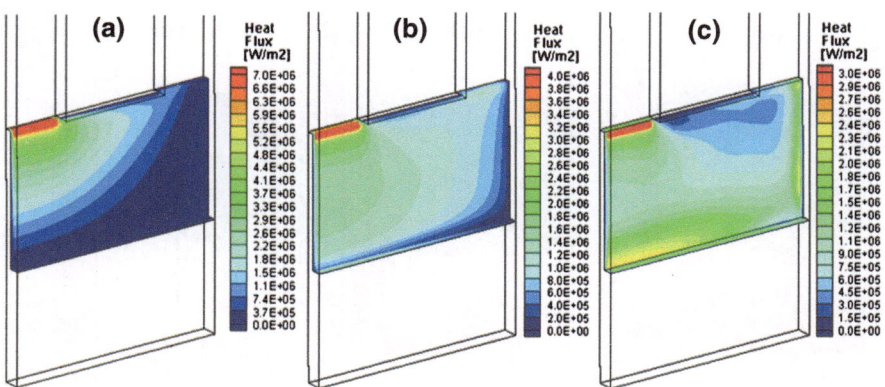

Fig. 2.12 Fin feat flux distribution for **a** Case #1 at $h = 50,000$ W/m²K; **b** Case #2 at $h = 150,000$ W/m²K; **c** Case #3 at $h = 250,000$ W/m²K (Cetegen 2010)

channel. In this region, the flow turns toward the outlet of the microchannel and does not flow deeply into the corner of the channel, which has much higher flow resistance. The third recirculation zone appears just after the outlet from the microchannel, in the outlet feed channel. Similarly, the velocity distribution tends to be stratified, with the high velocity core close to the center of the outlet feed channel. This separation is the result of the high outlet velocities at the sharp manifold corner.

In addition to the flow conditions discussed above, the flow velocity can affect the flow field by creating secondary flows. Due to the bulk flow rotation, a centrifugal force acting from the center of rotation to the channel bottom becomes present, and the fluid at the side walls is pressed in the opposite direction. This, in turn, forces the fluid to generate vortex pairs that can fill the channel cross-section. The magnitude of the centrifugal force depends on the flow velocity and radius of curvature of the bend and is a function of the Dean number. The development of secondary flows changes both heat transfer and flow characteristics. The vortices create continuous fluid mixing by moving the colder fluid at the center to the side walls of the microchannel, thus enhancing the convective heat transfer. This process disturbs and reduces the thickness of the thermal boundary layer that develops at the inlet to the microchannel. On the pressure drop side, however, due to the additional flow energy associated with vortices, the pumping power required to drive the fluid in a curved pipe will be always higher than straight tube and channel geometries at the same flow rate. These conflicting objectives of heat transfer efficiency and pumping power need to be considered to clearly evaluate the possible benefits of secondary flows.

Velocity vectors at several cross-sections in the computational domain for Case #3 are shown in Fig. 2.9. As the fluid starts to turn at both the first and second bends, vortex pairs are created in the microchannel as the fluid in the center is pushed down and forces the liquid close to the walls to move in the opposite direction. Due to symmetry conditions, only one half of the channel,

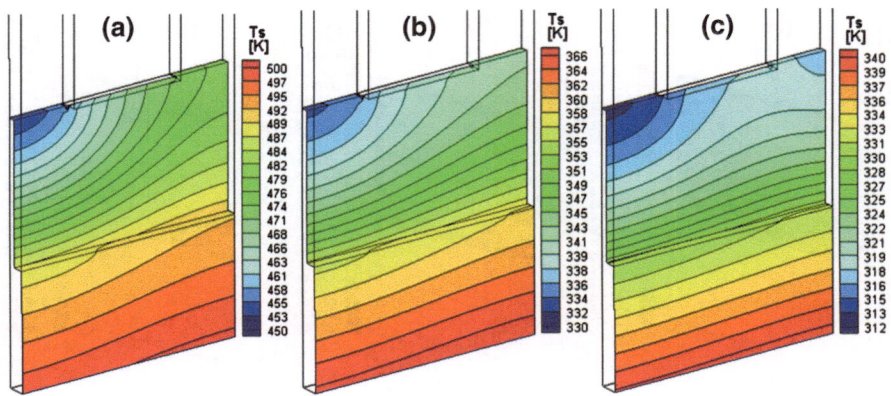

Fig. 2.13 Temperature contours of fin surface, microchannel bottom wall and base material for **a** Case #1 at $h = 50{,}000$ W/m²K; **b** Case #2 at $h = 150{,}000$ W/m²K; **c** Case #3 at $h = 250{,}000$ W/m²K (Cetegen 2010)

and therefore only one vortex, is shown in the figure. The pathlines and transition between flow regimes at different Reynolds number flows are shown in Fig. 2.10a, b and c. The pathlines are uniform and almost symmetric, with no disturbance in the flow. As the Reynolds number is increased in Case #2 and later in Case #3, the centrifugal forces start to become important. The vortices formed during flow turning disturb the flow, and the pathlines show a less uniform pattern. As seen in Fig. 2.10c, due to secondary flows and the mixing effect, the pathlines can cross over each other on the plane of view.

The fluid temperature distribution at the symmetry plane of $z = 0$ is shown in Fig. 2.11a, b and c for Case #1, Case #2 and Case #3, respectively. Two different trends become important as the flow Reynolds number goes from low to high. First, the temperature rise of the fluid is significant at low flow rates. This is expected, since the mass flux is lower and the fluid can be heated much more before exiting the microchannel. Second, the velocity stratification shown in Fig. 2.7 also creates temperature stratification. The low velocity zones close to the bottom of the microchannel at low Reynolds number flows and "dead zones" in the high Reynolds number cases create high temperature zones. Similarly, the relatively low velocities allow the liquid to be heated for longer times, and therefore a temperature difference is observed in these regions.

The heat flux distribution on the fin and microchannel base for the three cases is shown in Fig. 2.12a, b and c. Although the mass flux has a significant effect on heat flux distribution on the fin surface, the maximum heat flux always occurs at the inlet region to the microchannel. This global maximum is the result of the large temperature difference between the fluid and surface and the significant entrance region effects where the thermal boundary layer is relatively thin and heat transfer resistance is low.

At low Reynolds number flows (Fig. 2.12a) the heat transfer efficiency decreases due to increase in noneffective areas that do not contribute to heat

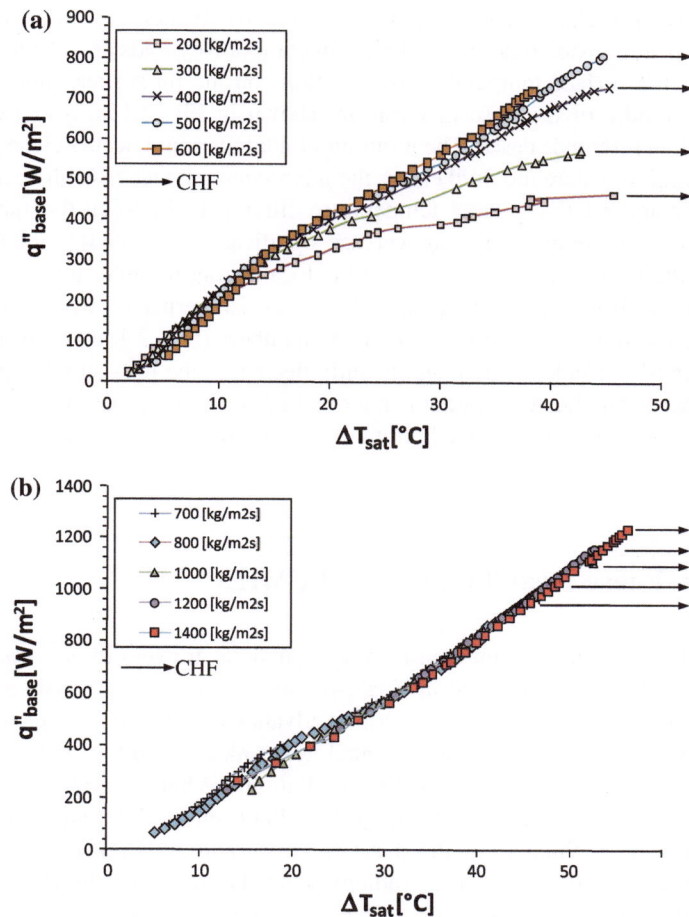

Fig. 2.14 Boiling curves for R-245fa at **a** $200 < G < 600$ kg/m^{2}s and **b** $700 < G < 1400$ kg/m^{2}s (Cetegen 2010)

transfer. The temperature difference between the bulk liquid and surface is low in these regions. The heat transfer characteristic changes significantly with higher Reynolds number flows (Fig. 2.12c). First, the heat flux becomes more uniformly distributed along the fin surface with the heat flux gradient decreasing slightly along the microchannel length. However, local minimums are present at the heat transfer surfaces under the recirculation zones. The fluid trapped in the recirculation zones has high bulk temperatures, which decrease the convective heat transfer efficiency. Second, the secondary local maxima starts to appear close to the impingement zones under the inlet and outlet feed channels. Similar local high heat transfer zones were reported by Copeland (1995), Copeland et al. (1997) and Ng and Poh (1999). Close to the impingement zones, the local fluid acceleration and vortices created during the flow turning have the effect of reducing the thermal boundary layer thickness and enhancing heat transfer. It should be noted that these local high heat transfer zones are a strong function of Reynolds and Dean numbers.

Finally, the surface temperature distributions on the fin surface, microchannel base, and middle of the base material are shown in Fig. 2.13a, b and c for the selected three reference cases. The minimum fin temperature occurs close to the tip of the fin region, where the fluid enters the microchannel. Due to high heat transfer coefficients and relatively large temperature differences between the surface and fluid, the local fin temperature can decrease significantly. At low Reynolds number flows, all the heat transferred to the fluid is localized at the inlet region with heat flowing from all directions, perpendicular to the isotherms shown in Fig. 2.13a. Increasing the flow rate to higher Reynolds numbers (Fig. 2.13a and b) increases the heat transfer efficiency and significantly decreases the heat sink temperatures. On the other hand, the temperature isotherms become more horizontal, indicating a more uniform heat flow from the bottom of the base material to the fin heat transfer surface.

2.4 Two-Phase Heat Transfer in FFMHX

The FFHT flow configuration can be very beneficial for two-phase heat transfer applications. First, as discussed in single-phase flow, the overall system pressure drop decreases dramatically. This is a key advantage of FFMHX over traditional cooling technologies such as microchannel heat sinks, where the cooling capacity is mainly dictated by the pressure drop limitations. When working in two-phase heat transfer mode, large pressure differences between the inlet and outlet of the heat sink create significantly high saturation temperature differences. This, in turn, creates an unwanted temperature gradient along the flow and eliminates the isothermal flow conditions. Second, as the system pressure drop is lowered, the system pumping power is also decreased. This is also a key parameter for systems running continuously for long periods of time. Third, as the microchannel lengths become shorter and shorter, two-phase flow tends to become more stable. For example, for the same flow rate and outlet qualities, a shorter tube can experience much higher critical heat flux values when compared with a longer one. Also, when the microchannel length is comparable with its hydraulic diameter, the bubble dynamics observed in microchannels are believed to become more stable. Here, the fast growing bubble can reach the channel outlet faster than it can in traditional long microchannels.

Cetegen (2010) experimentally investigated force-fed flow boiling using copper microgrooved surfaces and a refrigerant, R-245fa, as the working fluid. The fins of the microgrooved surface were 480 μm high and 85 μm wide, while the microchannel gap was 42 μm. The experimental tests were performed at 10 different mass flux conditions ranging from 200 to 1,400 kg/m^{2}s, as shown in Fig. 2.14a and b. The boiling curves are plotted for the base heat flux q''_{base} versus wall superheat ΔT_{sat}, which is the temperature difference between

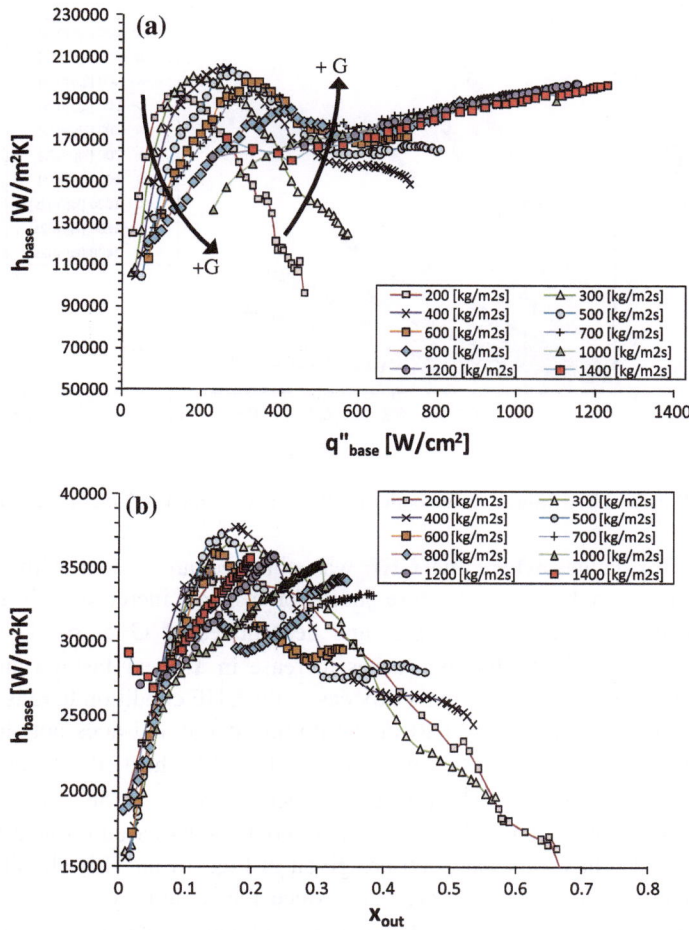

Fig. 2.15 a Heat transfer coefficient based on base area versus base heat flux and **b** heat transfer coefficient based on wetted area versus outlet quality for refrigerant R-245fa (Cetegen 2010)

average base surface temperature and average saturation temperature. The trend of the curves can be summarized as follows. For low to medium heat fluxes of $q''_{base} < 300 \, W/cm^2$, the boiling curves show an almost linearly increasing trend with slightly different slopes. In this region, the wall superheat increases when going to higher mass fluxes at a constant heat flux. This implies higher heat transfer efficiency at lower mass fluxes. On the other hand, when heat fluxes increase above $q''_{base} > 300 \, W/cm^2$, the trend is reversed, and departures from the main trend are observed for low to medium mass fluxes of $200 < G < 600 \, kg/m^2 s$. The departing curves have a smaller slope on the chart, which indicates less heat transfer for a given wall superheat; therefore, the heat transfer efficiency is expected to decrease compared with the main trend. The slope will continue to decrease until it reaches CHF conditions.

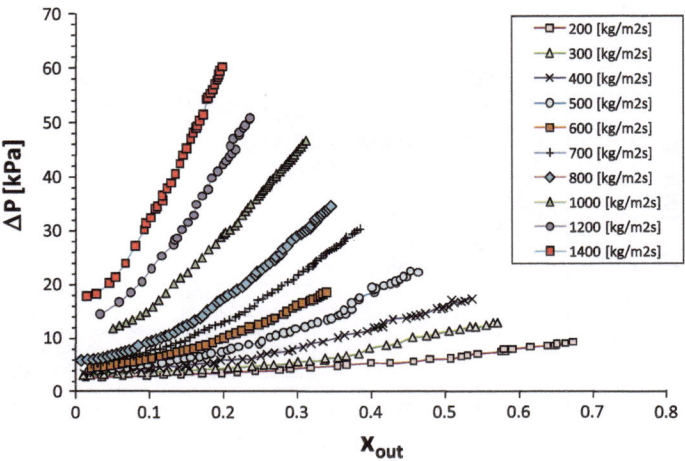

Fig. 2.16 Pressure drop values versus outlet quality for refrigerant R-245fa (Cetegen 2010)

The heat flux values where the CHF was obtained are shown with the black arrow, denoting a large temperature jump for a small increase in heat flux.

For medium to high mass fluxes and heat fluxes of $G > 600 \, \text{kg/m}^2\text{s}$ and $q''_{\text{base}} > 600 \, \text{W/cm}^2$, the boiling curves increase in a linear fashion and almost overlap. In contrast to the low mass flux cases, the CHF condition here is achieved along the straight overlapping part of the boiling curve and does not show early performance degradation. The maximum achievable heat flux was $q''_{\text{base}} = 1.23 \, \text{kW/cm}^2$, corresponding to a wall superheat of $\Delta T_{\text{sat}} = 56.2 \, ^\circ\text{C}$.

For the same set of data, the heat transfer coefficients based on heat sink base area versus base heat flux variation are given in Fig. 2.15a. This classification is more convenient for heat sink designers since the definitions are based on the targeted cooling area and are not dependent on enhanced area parameters such as channel aspect ratio or fin efficiency. Figure 2.15b shows the heat transfer coefficients based on wet channel area versus outlet quality. This definition is useful for comparing the hydraulic and thermal performances of different microchannel geometries of microgrooved surfaces. It is important to note that these two graphs are not independent because for a fixed heat flux and mass flux there exists only one outlet quality. In other words, for a constant mass flux case, to change the outlet equilibrium quality the heat flux needs to be changed and vice versa.

As shown in Fig. 2.15a, at mass fluxes below $G < 500 \, \text{kg/m}^2\text{s}$ the heat transfer coefficient curves show a bell-like curve starting with an initially increasing trend and then decreasing gradually until the CHF occurs. Here, the increase in heat transfer coefficients for all mass fluxes has a similar slope, while the decreasing trend is more dependent on mass flux. The increase in mass flux decreases the slope of the decreasing part of the curve, indicating an improvement in heat transfer and less severe performance degradation. The maximum peak point remains between $190,000 < h_{\text{base}} < 200,000 \, \text{W/m}^2\text{K}$ and shifts toward higher

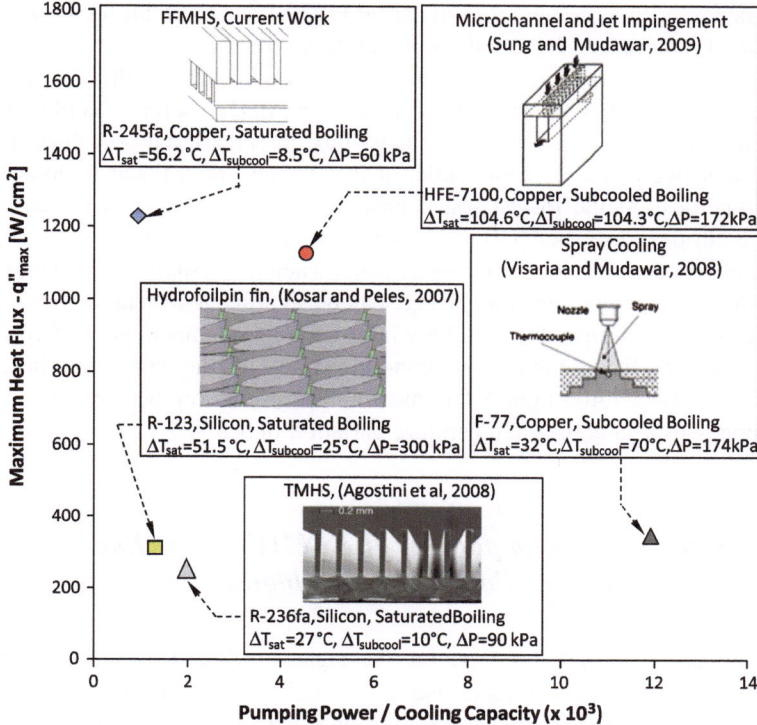

Fig. 2.17 Thermal performance comparison of different high heat flux cooling technologies (Cetegen 2010)

heat fluxes with increasing mass flux. It is interesting to note from Fig. 2.6b that the maximum heat transfer coefficients at these flow rates remain almost constant at $x_{out} \approx 0.15$. Here, the steep increasing trend of heat transfer coefficients may suggest a flow regime dominated mostly by nucleate boiling, which is a function of surface heat flux. The decreasing trend observed after the maximum may be the result of local dryouts, which become less severe as the mass flux increases.

For heat transfer coefficients obtained for higher mass fluxes of $G > 500$ kg/m^{2}s, the trend shifts, as seen in Fig. 2.15a. As the mass flux increases, the peak previously seen at lower mass fluxes further decreases and diminishes after $G = 1,000$ kg/m^{2}s. At the same time, the previously decreasing trend is also eliminated, and for high mass fluxes, both effects create a slow but monotonically increasing trend. More interestingly, all curves collapse altogether, forming a single line on the graph. This trend may suggest a convective boiling-dominated heat transfer regime, since the heat transfer coefficients are slightly dependent on heat flux but are more dependent on outlet quality and mass flux (Fig. 2.15b). Another observation from these graphs is the trend of heat transfer coefficient before reaching CHF. For low mass fluxes and the bell-like curve trend, the CHF always occurs when the heat transfer coefficients have a decreasing trend. On the other hand, at high mass fluxes, the boiling crisis phenomenon occurs on the curve shown in Fig. 2.15a, where all the heat transfer coefficients overlap and have a slightly increasing trend.

The pressure drop variation versus outlet equilibrium quality is shown in Fig. 2.16. As expected, the pressure drop of the FFMHX is a function of both mass flux and outlet quality. For a constant mass flux, the pressure drop values show an exponential-like increase at low outlet quality values and a linear trend at higher outlet qualities. At the lowest tested mass flux of $G = 200$ kg/m^{2}s, the system pressure drop is less than 10 kPa, corresponding to a saturation temperature change of less than 1.5 °C. On the other hand, when the system works at high mass fluxes the saturation temperature change can be significant and should be carefully considered in the design stages. A very large saturation temperature difference can increase the surface temperature non-uniformity of the heat sink base. For example, for $G = 1,400$ kg/m^{2}s the maximum pressure drop was 60.4 kPa, which corresponds to a saturation temperature variation of 7.8 °C.

2.4.1 Performance Comparison of FFMHX with Other High Heat Flux Cooling Technologies

As mentioned in the previous section, the highest critical heat flux obtained for force-fed two-phase heat transfer was $q'' = 1230$ W/cm^2 measured for average wall superheat of 56.2 °C and subcooling of 8.5 °C. The corresponding pressure drop was measured as 60.3 kPa, and pumping power was calculated as 1.13 W. The heat sink footprint area was 7.8×7.8 mm^2. As a general design rule, a heat sink should perform at high heat fluxes with low wall superheat, low subcooling, low pressure drop, and low pumping power. Therefore, compared to single-phase heat transfer, it is more challenging to compare cooling technologies for two-phase heat transfer mode because the heat sink performance is dependent on many more parameters. Nevertheless, a quantitative comparison can still be made by plotting the data over the two most important parameters. For this purpose, the two comparison parameters were selected as maximum heat flux and pumping power over cooling capacity ratio. For these parameters, the performance of FFHT was compared with other competing high heat flux cooling technologies (Sung and Mudawar 2009; Visaria and Mudawar 2008; Kosar and Peles 2007; Agostini et al. 2008), and the resulting graph is shown in Fig. 2.17. It should be noted, however, that the comparison parameters were selected based on their importance to the objective of the study. For example, for a comparison based on heat transfer efficiency of the y axis can be replaced with heat transfer coefficient h.

The technology comparison shown in Fig. 2.17 clearly indicates that the FFMHX technology can remove much higher heat fluxes (much higher heat transfer coefficients) with substantially less pumping power requirements than any of the applicable cooling technologies investigated. More importantly, this is accomplished with a reasonable wall superheat, with low or little subcooling and moderate pressure drops. Another important point for the comparison chart is that no two-phase optimization was reported for any of the heat sink designs. For optimum conditions, the sequence and location of points on the graph may be significantly different. Therefore, future comparisons of the technologies should consider the effect of optimum design on thermal performance.

Chapter 3
Emerging Applications of Microchannels

Abstract Microchannel heat exchangers have applications in several important and diverse fields including: aerospace; automotive; bioengineering; cooling of gas turbine blades, power and process industries; refrigeration and air conditioning; infrared detectors and powerful laser mirrors and superconductors; microelectronics; and thermal control of film deposition. The advantages of microchannel heat exchangers include high volumetric heat flux, compactness for space-critical applications, robust design, effective flow distribution, and modest pressure drops. This chapter will cover selected industrial examples for microchannel heat exchangers, microchannel heat pipes, and microchannel heat plates.

Keywords Automotive · Aerospace · Chemical reactor · Cryosurgery · Laser diode · Heat pipe · Pulsating · Cosmos · Heat plate

3.1 Microchannel Heat Exchangers

3.1.1 Automotive and Aerospace

Microchannel heat exchangers have at least one fluid flow passage with typical dimensions between 1 μm and 1 mm and have great potential in process intensification of various industrial areas (Fan and Luo 2008). There are many possible channel geometries for microchannel heat exchangers, two types of which are the most widely used in compact heat exchanger designs for automotive and aerospace applications. These are shown in Figs. 3.1 and 3.2. High-temperature and compact micro heat exchangers can be manufactured using ceramic tape technology (Schmitt et al. 2005), which uses fused ceramic layers to create channels with dimensions below 1 mm (Ponyavin et al. 2008). Metal-based microchannel heat

M. Ohadi et al., *Next Generation Microchannel Heat Exchangers*,
SpringerBriefs in Thermal Engineering and Applied Science,
DOI: 10.1007/978-1-4614-0779-9_3, © The Author(s) 2013

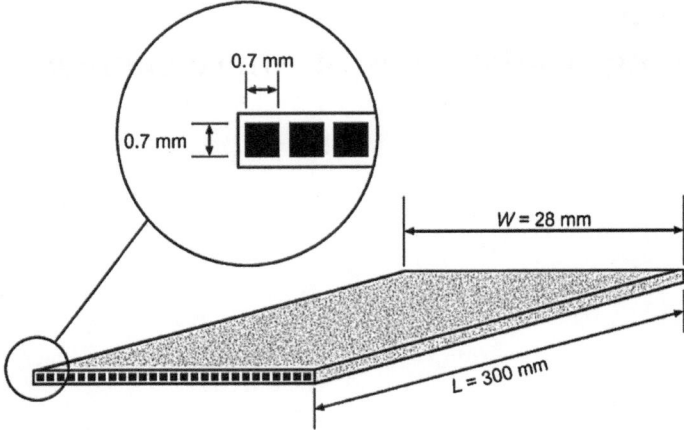

Fig. 3.1 Typical dimensions of a rectangular microchannel for compact heat exchanger applications

exchangers are also of current interest because of the combination of high heat transfer performance and improved mechanical integrity (Mei et al. 2008). Figure 3.3 shows a manufactured, flat extruded multichannel aluminum heat exchanger.

With the aim of reducing size and cost, microchannel heat exchangers are now able to achieve performances for surface area per unit volume as high as 1500 m^2/m^3. Their fin geometry is rather complex (as shown in Fig. 3.4), as they are specially designed to augment the heat transfer level on both the liquid and air sides with a balanced resistance between the two sides. Several experimental correlations for compact heat exchangers are available, but their current technical limitations may not allow for the practical design and optimization of new microchannel heat exchangers.

In recent years major progress in microchannel heat exchangers has been made by the automotive, aerospace, chemical reactor, and cryogenic industries. Thermal duty and energy efficiency requirements have increased during this period, while space constraints have become more restrictive. The trend has been toward greater heat transfer rates per unit volume. The hot side of the evaporators in these applications is generally air, gas, or a condensing vapor. With advances in the air-side fin geometry, heat transfer coefficients have increased, as have surface area densities. As the air-side heat transfer resistance has decreased, more aggressive heat transfer designs have been sought on the evaporating side, resulting in the use of microchannel flow passages on the liquid side (evaporating or condensing or single-phase regimes). The major changes in recent evaporator and condenser designs for automotive and other compact heat exchanger applications involve the use of individual, small-hydraulic-diameter flow passages arranged in a multi-channel configuration on the liquid side.

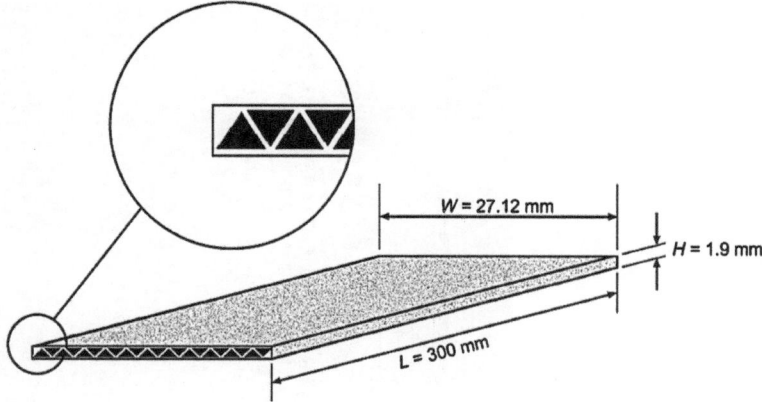

Fig. 3.2 Typical dimensions of a triangular microchannel for compact heat exchanger applications

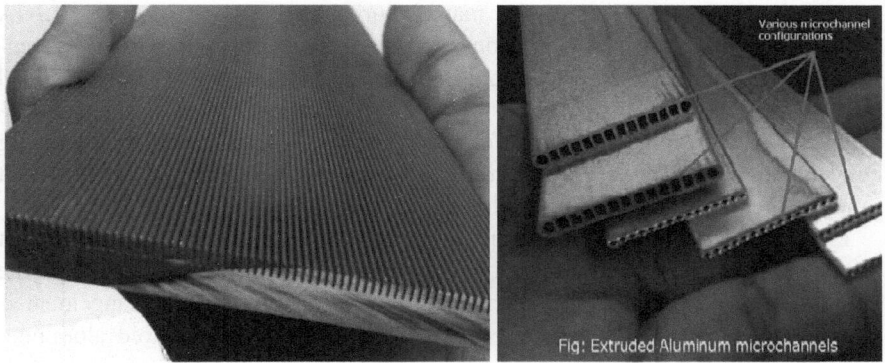

Fig. 3.3 View of a microchannel heat exchanger

The ability to efficiently transfer heat between fluids using lightweight, compact heat exchangers is important in a variety of applications, such as automobile radiators, air conditioning, and aerospace applications. Microchannel heat exchangers are well suited to these applications due to the microchannels' compactness, lightness, and high heat transfer performance.

Car radiators employ a cross flow design that allows a sufficient mass flow rate of air through the radiator while using only the stagnation pressure associated with the motion of the automobile. The common measure of performance for car radiators is the heat transfer/frontal area normalized by the difference in inlet temperatures of the coolant (water–glycol) and the air. For conventional car radiators, 0.31 W/Kcm^2 of heat transfer/frontal area can be obtained between the air and the coolant (Webb and Farrell 1990; Parrino et al. 1994). However, these radiators are extremely thick (1–2 cm) compared with the thickness of the micro heat exchanger described here (0.1–0.2 cm). Harris et al. (2000) designed and

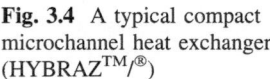

Fig. 3.4 A typical compact microchannel heat exchanger (HYBRAZ^{TM}/$^{®}$)

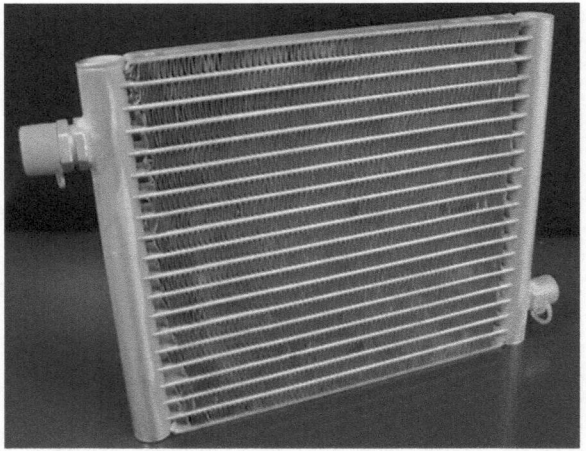

fabricated a cross flow micro heat exchanger that transfers heat from coolant (water–glycol) to air. They used a manufacturing process that combines LIGA micromachining, traditional precision micromachining, and bonding. They compared the thermal performances of plastic, ceramic, and aluminum microchannnel heat exchangers with those of conventional car radiators. The cross flow microchannel heat exchanger can transfer more heat/volume or mass than existing conventional heat exchangers within the design constraints. This can be important in a wide range of applications (automotive, home heating, and aerospace).

Figure 3.5 shows a plate-fin evaporator geometry commonly seen in compact refrigerant evaporators for automotive and aerospace applications. As seen in the figure, fins are placed between microchannel flow slabs, and the arrangement is brazed together in a special oven. Figure 3.6 shows a manufactured aluminum microchannel heat exchanger.

3.1.2 Chemical Reactors

Microchannel chemical processing technology is an emerging field with applications in most industrial processes due to its excellent mixing capabilities, controlled reaction environment, and energy efficiency. This technology offers improvements in existing processes and will enable new processes to become cost effective. The basic microchannel reactor design is based on the flow between parallel platelets coated with a catalyst. The large aspect ratio of the channel provides extensive surface area in a small volume. Microchannel reactors have been developed based on ceramic substrates as well as metal substrates. In both types of reactors, multiple layers coated with catalytic material are bonded, forming a monolithic structure. An added benefit of a layered pattern is the ability to easily scale up or down by adjusting the number of layers. This provides great

Fig. 3.5 Photographs of refrigerant and airside flow passages in a compact automotive microchannel heat exchanger. **a** A compact microchannel heat exchanger for automotive applications (Flygrow Refrigeration Co., Ltd); **b** air-side fin

(a)

(b)

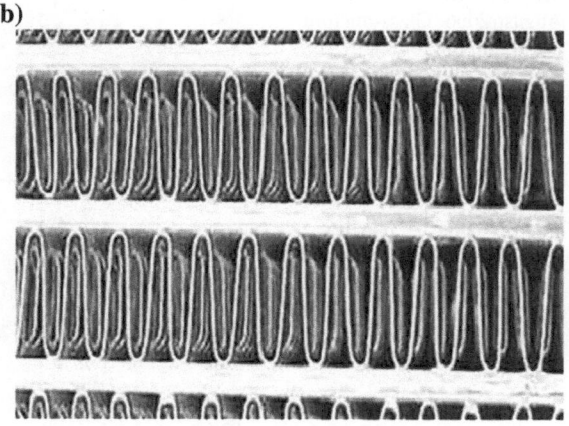

Fig. 3.6 Aluminum microchannel heat exchanger (Alarko-Carrier Inc.)

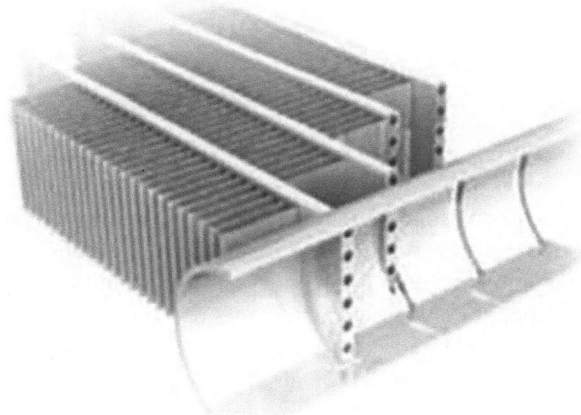

flexibility in the design, since if the production capacity needs to be changed, there is no need to redesign the reactor. Figure 3.7 shows a manufactured microreactor.

Recently, microchannel heat exchangers have been widely used in fuel cell systems, which have gained increased attention in the energy system sector.

Fig. 3.7 Glass microreactor: the channels of the chip in the picture are 150 μm wide and 150 μm deep (This image was published under a Creative Commons Attribution license and appears on Wikimedia Commons at http://en.wikipedia.org/wiki/Microreactor, Micronit, Glass Microreactor made by Micronit Microfluidics, 11 August 2006, Glass-microreactor-chip-micronit.jpg)

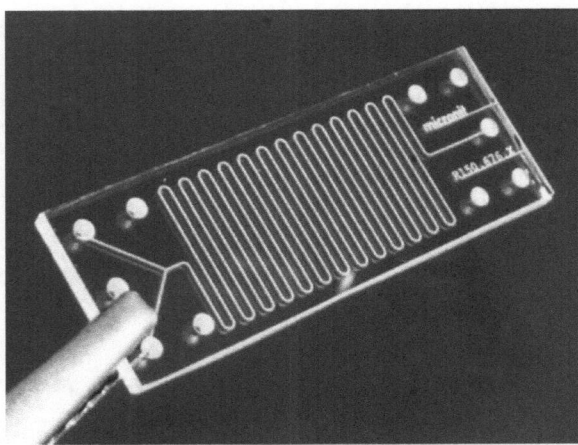

Fig. 3.8 Portable SOFC process flow diagram (Murphy et al. 2011)

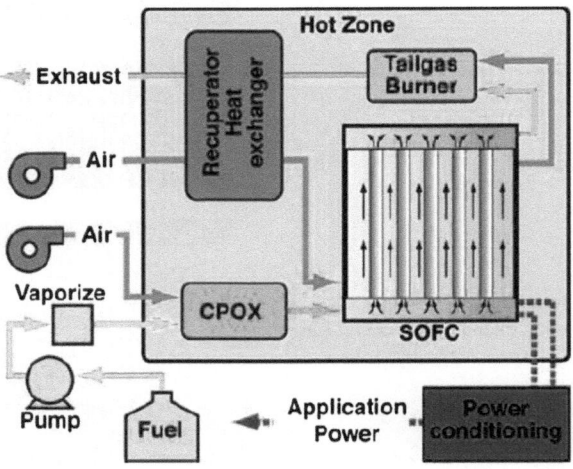

The thermal integration of heat exchangers and chemical reactors is essential for effective operation of solid oxide fuel cell (SOFC) systems. In Fig. 3.8, a partial-oxidation reactor, a catalytic tail-gas combustor, and a cathode-air preheating recuperator are shown as separate unit processes. However, system efficiency can be improved and costs reduced via process intensification, which combines unit processes. Figure 3.9a is an exploded view illustrating the internal geometry for two layers of microchannels, with a photograph of the microchannel reactor shown in Fig. 3.9b. The current reactors use four layers, two reacting and two non-reacting, and can be configured as required for different applications (such as steam reforming or catalytic partial oxidation). Microchannels within the reacting layers are washcoated with an Rh-catalyst for fuel reforming. The hot gases flowing through the non-reacting layers heat the fuel and drive the upstream fuel-reforming processes within the reactive channels. The current reactors are

Fig. 3.9 **a** Exploded illustration of two layers of the ceramic microchannel reactor, and **b** photograph (Murphy et al. 2011)

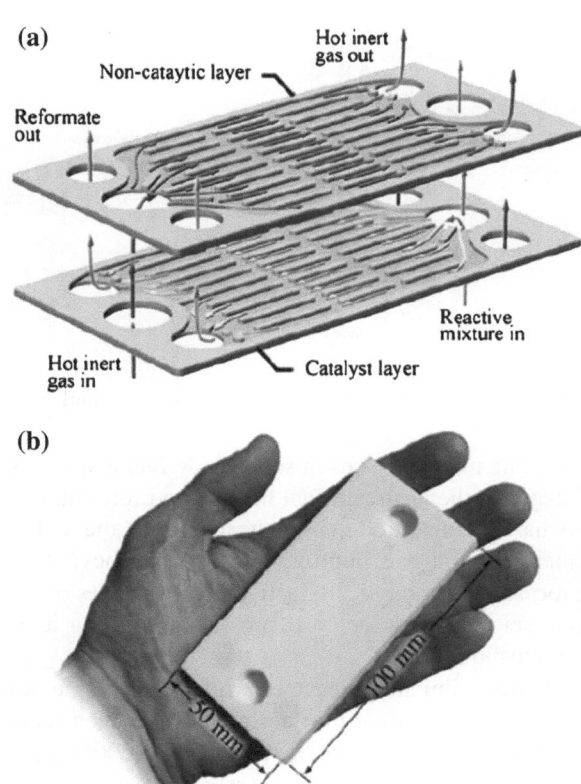

fabricated by CoorsTek, Inc. (Golden, Colorado, USA) using 94 % alumina. A new fabrication process (called Pressure Laminated Integrated Structures, or PLIS) enables cost-effective manufacturing.

The performance advantages of microchannel reactors, including integrated chemical processing, are well documented (Kandlikar et al. 2006; Yarin et al. 2009; Anxionnaz et al. 2008). The fine channels utilized in such reactors enable exceptional control of thermal processes, enabling tight regulation of chemical reactions for achieving desired operating conditions. This thermal control results in higher product yields and selectivity, optimal catalyst activity, longer catalyst lifetime, and higher overall reactor productivity.

A recent review by Sommers et al. (2010) discusses numerous applications that particularly benefit from ceramic heat exchangers. Ceramic materials offer potentially significant advantages compared to metal alternatives, including significantly higher operating temperatures, improved tolerance of harsh chemical environments, improved bonding with ceramic-based catalyst washcoats, and significant cost savings in materials and manufacturing methods. Although there are numerous possible applications, the effort reported here is motivated by the need for low-cost, high-performance fuel reformers, and the need to effectively

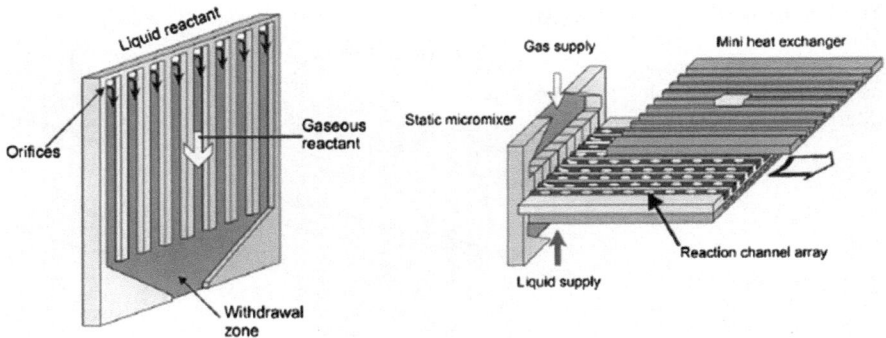

Fig. 3.10 Falling film and microbubble reactors (Jahnisch et al. 2000)

integrate unit processes in solid-oxide fuel cell systems. The microchannel reactor integrates the upstream fuel heating and reforming with downstream recuperation to harness the high-quality heat exiting the tail-gas combustor. Such process intensification can improve system efficiency. It is feasible to further combine processes by catalytically activating both sides of the microchannel reactor, using one set of layers for upstream fuel reforming and the opposite set for tail-gas combustion.

Falling-film microreactors work in the same way as macro-scale absorbers. Microchannel falling-film absorbers have been reported in the open literature by Goel and Goswami (2007), Hessel et al. (2005), Jahnisch et al. (2000), and Zhang et al. (2009). The central part of the microreactor is a stainless steel plate containing 64 vertically positioned microchannels (300 µm wide and 100 µm deep). Liquid spreads to form a thin film among the microchannels and flows further downward to the withdrawal zone at the bottom. Gas flows in a large gas chamber positioned above the microchannel section, facilitating concurrent or countercurrent operation mode. Figure 3.10 shows a schematic of a falling-film and microbubble reactor.

In membrane separation technology, gas and liquid flow in two different channels, separated by some physical structure, such as a mesh, selective membrane, or micro-porous plate (Abdallah et al. 2004). Oak Ridge National Laboratory and Velocys Inc. separated liquid and vapor phases in microchannels by membrane in a traditional counter-flow arrangement. The membrane prevents the liquid from bridging the microchannel and maintains stable phase separation. However, it significantly decreases the surface area of the liquid–gas interface. Figure 3.11 shows the opening area of the membrane comprising just about 10 % of total surface area.

In the co-current configuration, gas, and liquid flow concurrently in the same microchannel, creating various types of flow patterns similar to that of two-phase flow in a tube. The liquid film is maintained by shear stress of the moving gas phase, and film thickness can be ultimately small. Channel hydraulic diameter can be significantly smaller, limited only by the desired fouling characteristics.

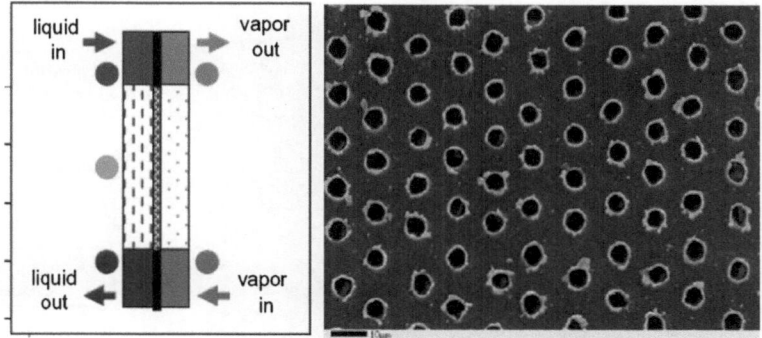

Fig. 3.11 Counter-flow absorber and phase separating membrane (Velosys Inc.)

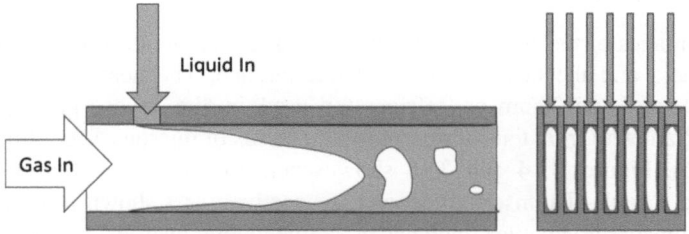

Fig. 3.12 Co-current microchannel absorption process (Jenks and Narayanan 2008)

With co-current flow, films of a few microns thick can be arranged on the channel wall, and associated with this thickness very high mass transfer coefficients can be obtained (Jenks and Narayanan 2008). Figure 3.12 shows a schematic of the co-current microchannel absorption process.

Yue et al. (2008) experimentally investigated hydrodynamics and mass transfer characteristics in co-current gas–liquid flow (CO_2/water) through a horizontal rectangular microchannel with a hydraulic diameter of 667 μm. Liquid-side volumetric mass transfer coefficient and interfacial area in the present micro-channel were measured as high as 21 s^{-1} and 9000 m^2/m^3, respectively, which are at least one or two orders of magnitude higher than those in conventional gas–liquid contactors. This shows the great potential of gas–liquid micro-chemical systems for many industrially relevant gas–liquid mass transfer operations and reactions.

3.1.3 Cryogenic Systems

Microchannel heat exchangers are used in cryosurgical probes for ablating tumors or treating heart arrhythmia (Marquardt et al. 1998), as shown in Fig. 3.13. The Joule–Thomson cryocooler is most widely used for the cryosurgical probe.

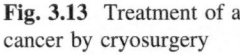

Fig. 3.13 Treatment of a cancer by cryosurgery

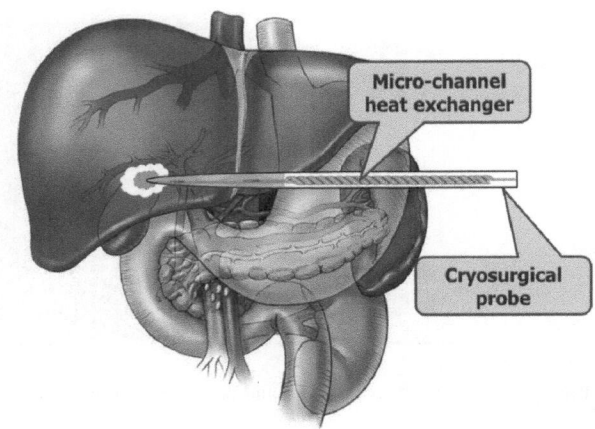

The advantages of the Joule–Thomson cryocooler include its simple structure and compactness and the lack of electrical interference. The heat exchanger of the subminiature Joule–Thomson refrigerator used in the cryosurgical probe has channel diameters from tens of microns to hundreds of microns, according to space limitations. Figures 3.14 and 3.15 show schematic diagrams of a cryosurgical probe and a Joule–Thomson cryocooler, respectively. As shown in Fig. 3.15, the Joule–Thomson cryocooler includes a counter-flow heat exchanger, a Joule–Thomson valve to decrease refrigerant temperature through an isenthalpic expansion process, and an evaporator to absorb external heat. Figure 3.16 shows a cryosurgical system for use in cancer surgery.

Until recently, the subminiature Joule–Thomson refrigerators in most commercialized cryosurgical probes have used a high pressure argon gas as the refrigerant in an open system. However, a closed system with a mixed refrigerant is desirable, since the efficiency of the Joule–Thomson refrigerator with a pure refrigerant is much lower than that possible with a mixed refrigerant. A closed-system cryosurgical probe with a mixed refrigerant is feasible because a mixed refrigerant decreases the working pressure of the refrigerator. The most important parameter to optimize in a Joule–Thomson refrigerator using a mixed refrigerant is the combination of the mixed refrigerant for high and low pressure conditions. In particular, pressure drop within the microchannel heat exchanger must be optimized to obtain optimal performance from the refrigerator.

Plants that liquefy natural gas are another application field of microchannel heat exchangers. Natural gas is liquefied at temperatures around -161 °C (112 K) (Waldmann 2008). The composition of natural gas depends on the well from which it is extracted, but most natural gas consists of methane (CH_4). Both global and domestic demand for liquefied natural gas (LNG), with its advantages for transportation and energy density, are currently increasing. LNG was the source of 25 % of world energy consumption in 2005, but this demand will increase up to 50 % in the next 10 years. Because 20 % of natural gas wells are under the sea, offshore gas plant development is being re-evaluated as a promising energy

Fig. 3.14 Schematic diagram of a cryosurgical probe

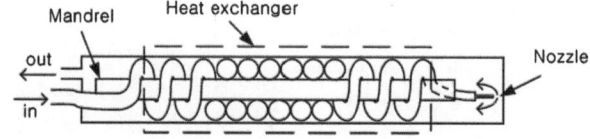

Fig. 3.15 Schematic diagram of a Joule–Thomson cryocooler

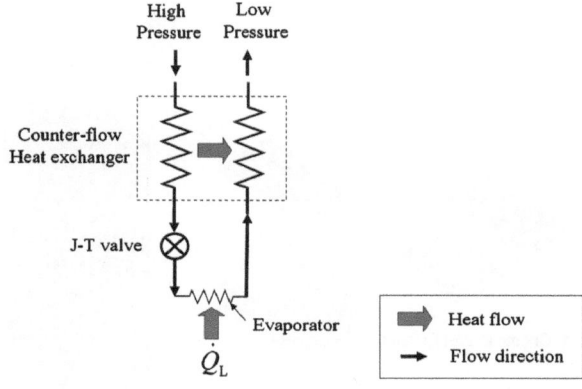

Fig. 3.16 Cryosurgical system for surgery of cancer (This image was published under a Creative Commons Attribution license and appears on Wikimedia Commons at http:// en.wikipedia.org/wiki/ File:Cryogun.jpg, Warfieldian, Medical cryotherapy gun used to treat skin lesions, 18 July 2011)

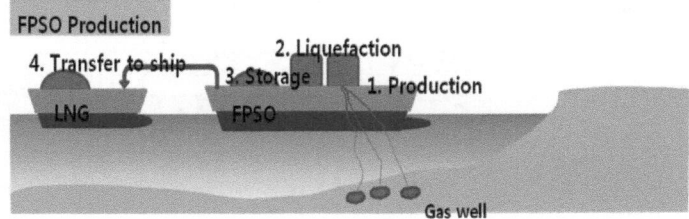

Fig. 3.17 Floating production storage and offloading method (Baek et al. 2010)

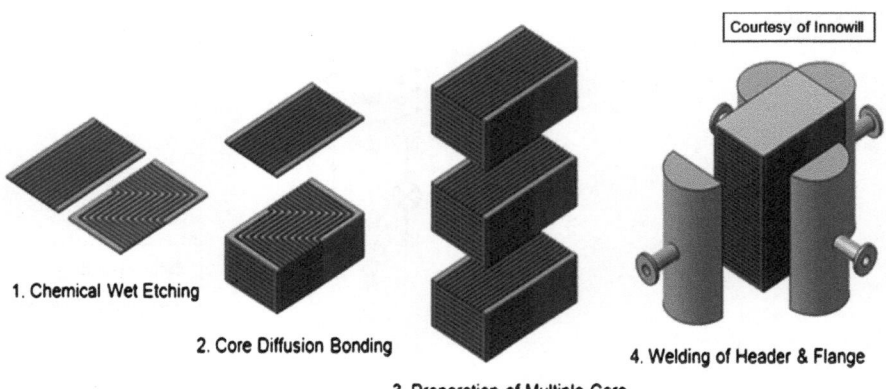

Fig. 3.18 Microchannel heat exchanger for FPSO (Baek et al. 2010)

industry. Floating production storage and offloading (FPSO) plants refine, liquefy and store natural gas from the sea. This LNG production method has a big advantage in that there is no need to transfer the natural gas from an offshore drilling rig to an onshore plant to liquefy it. An FPSO plant can also transport LNG to another ship, which is cost effective. The FPSO method is shown in Fig. 3.17.

There are some technical constraints (lightness of the ship, compactness, safety) for FPSO plants because many types of equipment must be loaded on the comparatively small ship. However, compactness and lightness can be achieved by using the proper equipment. One of the largest pieces of equipment in an LNG plant is the heat exchanger, and thus, if the size of the heat exchanger can be reduced, the size of the entire facility can be reduced. Microchannel heat exchangers can be used to achieve this goal. A compact heat exchanger for FPSO constructed by the etching method and diffusion bonding is shown in Fig. 3.18. A manufactured microchannel heat exchanger for FPSO is shown in Fig. 3.19. Microchannel heat exchangers for FPSO have the advantage of compactness, but they also have a disadvantage in performance due to axial conduction, especially in cryogenic applications (Maranzana et al. 2004). In many cryogenic applications, the desired heat transfer, which is the refrigeration load, is a small fraction of the heat transferred within the heat exchanger. In this case, the ineffectiveness of the

Fig. 3.19 Microchannel heat exchanger for an FPSO (Kim et al. 2010)

heat exchanger must be an even smaller number, and the effect of axial heat conduction and parasitic heat transfer can dominate the performance of the device (Nellis 2003). Furthermore, the large absolute temperature change within the heat exchanger produces correspondingly large variations in the properties of the fluids and metal that can also affect performance.

3.1.4 Laser Diode Applications

Microchannel cooling is now a mature technology, over 30 years old and widely used on a commercial basis in the high-power laser diode industry. Microchannel coolers are also now being used in commercial systems to cool LEDs in UV curing systems and photovoltaic cells in concentrated solar power systems. The present commercial cooling technologies, however, will not be adequate as power levels of semiconductor devices increase. Since laser diodes have stringent performance requirements and are by far the largest commercial market share of microchannel coolers, the remainder of this discussion will focus on laser diode applications.

Commercial coolers generally employ water-cooled copper microchannels to dissipate up to 1 kW/cm^2 from semiconductor heat sources ranging from 0.1–2 cm^2 in size. These coolers use rectangular ducts with widths from

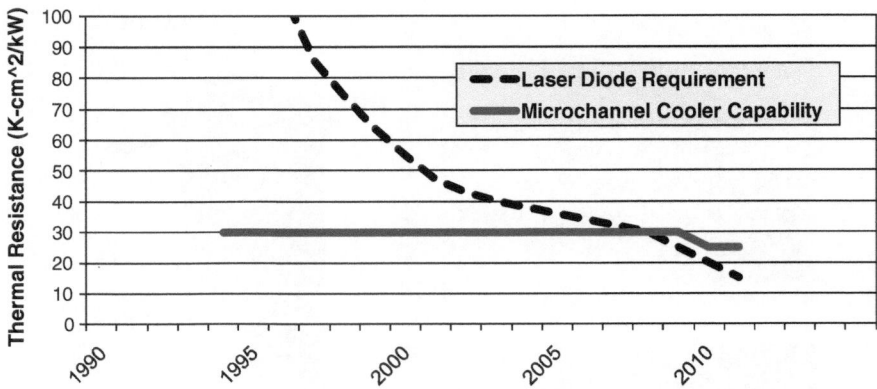

Fig. 3.20 Microchannel cooler capability versus laser diode requirements

25–200 μm, duct aspect ratios up to 15X, and flow rates ranging from 10–30 gm/cm^2/W (based on heated area). These parameters result in thermal resistances of 30–75 K-cm^2/kW, with pressure losses between 10 and 50 psi.

Laser diodes stress thermal management capabilities due to their low allowable junction temperatures (<70 °C) and large waste heat fluxes. Historically the capabilities of microchannel coolers have exceeded the most demanding laser diode requirements, accommodating heat fluxes up to 1 kW/cm^2 (Fig. 3.20). However, the operating power of laser diodes has continuously increased, and currently prototype systems with heat fluxes up to 2.5 kW/cm^2 are being considered, suggesting that for allowable temperature differentials of 30–50 °C, future microchannel coolers must be able to operate at thermal resistances from 12 to 20 K-cm^2/kW. Unless heat spreaders are employed, this is beyond existing commercial capability.

There are four main ways future commercial microchannel systems can be improved: (1) reducing the cooler thermal resistance; (2) reducing package thermal resistance through use of reduced thermal expansion coolers; (3) reducing flow rate to minimize system costs of cooling; and (4) increasing operating lifetime.

Reducing thermal resistance is critical to the continued increase in laser diode array brightness. Methods under consideration for this task include reducing channel widths, adding secondary features, such as fins or perforations, and using high conductivity materials such as CVD diamond. Reducing channel width is promising: for water-cooled systems, reducing the channel width of a copper microchannel cooler to ∼12.5 μm results in thermal resistances in the 15–20 K-cm^2/kW range with reasonable pressure losses. Secondary features can offer performance improvements for larger channels (channel width >100 μm), but they cannot be economically fabricated in smaller channels.

Diamond is well known for its outstanding thermal conductivity, but it is difficult to machine with micropassages. Figure 3.21 shows a prototype diamond microchannel cooler fabricated by laser etching thin (100 μm thick) diamond

Fig. 3.21 CVD diamond
microchannel cooler with
250 × 300 μm rectangular
ducts (Campbell et al. 2006)

sheets and brazing the sheets together. This approach was not able to create passages narrower than 250 μm.

Many laser diode packages employ fairly thick ceramic carriers (on the order of 250 μm) to accommodate the thermal expansion mismatch between the GaAs diode (CTE = 6.9 μm/m/K) and the copper cooler (CTE = 16.6 μm/m/K). These carriers have thermal resistances comparable to that of the cooler itself. Coolers fabricated using low expansion materials can reduce the thermal expansion significantly, with only small penalties in thermal resistance.

Figure 3.22 shows two versions of reduced expansion coolers. Microchannel coolers offer promise for cooling systems with large numbers of semiconductors, such as MW-class laser systems and radar arrays. The system-level coolant requirements are very large and represent a significant barrier to deployment of these systems. Refrigeration or two-phase cooling loops can reduce coolant usage, so long as the thermal resistance is not compromised. While promising two-phase microchannel research has been conducted over the past decade, two-phase systems remain immature. More work is required to implement research results in the marketplace.

As microchannel coolers have evolved from an exotic technology to a commonplace component, their required operating lifetime has increased from 3×10^3 to 10^4 h. In radar cooling applications the lifetime may be expected to be as high as 10^5 h. At the high flow rates required to achieve the lowest thermal resistances, erosion can limit the performance of a cooler to $\sim 10^3$ h. Corrosion is also a critical concern. Internal protective coatings are now being investigated to reduce susceptibility to erosion and corrosion.

3.1.5 Enhanced Micro-Grooved Tubular Evaporators for Waste Heat Recovery Applications

As much as 50 % of industrial energy is lost as waste heat in the form of energy in exhaust gas streams, cooling water and heat losses from hot equipment surfaces and heated products. To improve energy efficiency—and thereby reduce the

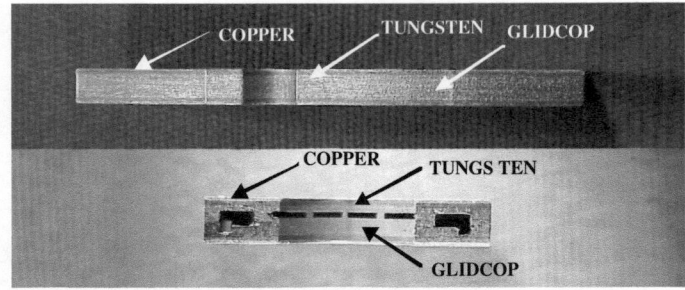

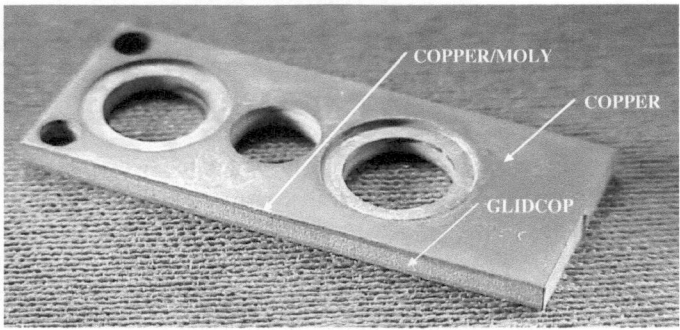

Fig. 3.22 Examples of reduced thermal expansion format microchannel coolers (Campbell et al. 2006)

emissions and cost associated with the energy resources—cost effective utilization of waste heat is of strong interest to many industrial sectors. Numerous technologies are commercially available for large-scale systems to tap waste heat and improve energy productivity. In some cases, waste heat recovery technologies have increased energy efficiency as much as 50 %. However, heat recovery is not economical for small-scale applications due to high capital costs. High system cost and heavy weight/volume of the associated equipment prevent the use of these technologies in smaller scale waste heat-to-cooling applications. This section describes the development of a high-performance, compact evaporator and condenser that will potentially reduce weight, volume, and cost of heat energy recovery as well as other energy conversion systems. The method uses an innovative microchannel technology combined with a force-fed fluid manifold system to enhance the heat and mass transfer in the evaporators and condensers.

Details of the novel force-fed microchannel technology used in this case are discussed in the literature (Baummer et al. 2008). Heat flux levels above 1100 W/cm^2 with h values above 200,000 W/m^2 K are reported using HFE7100 as the working fluid for electronics cooling applications. Compared to traditional microchannel design, very high heat transfer coefficients were obtained due to secondary flows and creation of evaporating thin liquid films. Also, the manifold system for guiding the flow reduced the length of the flow passes significantly and hence minimized the pressure drop. More recently, micro-structured surface

Fig. 3.23 **a** Schematic of the force-fed evaporation process; **b** tubular evaporator (Jha et al. 2011)

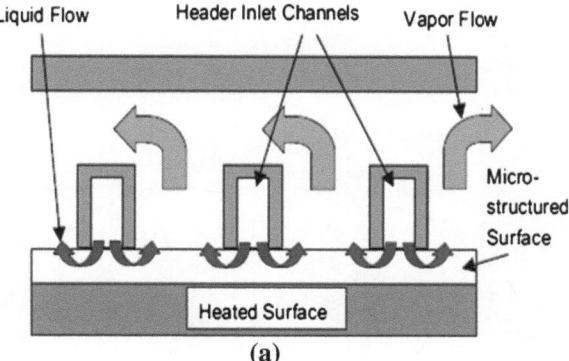

(a)

(b)

evaporators and condensers were repackaged to a tubular configuration to simplify fabrication and increase operation pressure. The first published information about tubular configuration is presented in (Jha et al. 2011).

A diagram of the force-fed evaporation process is shown in Fig. 3.23a. The process uses an inlet header positioned on top of the micro-structured surface to direct the working fluid into the surface. The working fluid evaporates as it passes through a heated surface, and vapor flows out of the channels back into the header and finally exits the header through alternate manifold channels. Figure 3.23b shows the fabricated tubular evaporator. The approximate weight of the evaporator is 1.3 kg, and it is roughly 60 mm in diameter and 250 mm in length.

A schematic of the experimental setup for investigating the heat transfer and pressure drop characteristics is shown in Fig. 3.24. The evaporator design was tubular and utilized microgrooves on the outer side and minigrooves on the inner side of the tubular surface for enhancing the heat transfer coefficient on both sides, thus improving the cooling capacity. The outer side is used for refrigerant (R134a)

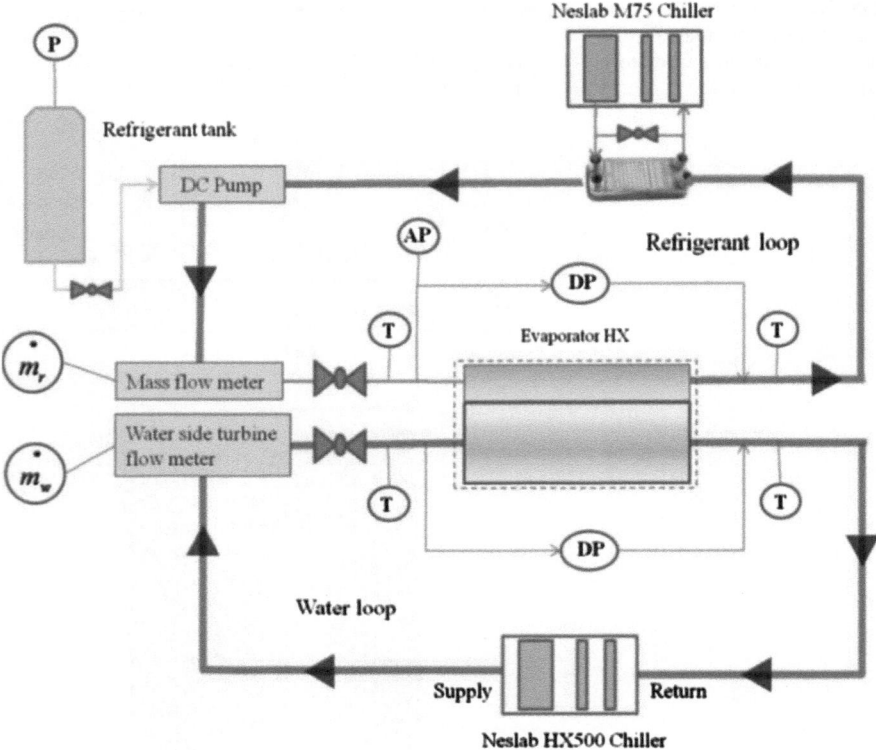

Fig. 3.24 Schematic of the experimental setup (Baummer et al. 2008)

flow, where medium aspect ratio microgrooves with hydraulic diameter of 150 μm
and inner side minigrooves with hydraulic diameter of 600 μm were used. The HX
tube was fabricated at Wolverine MicroCool Inc., utilizing patented micro-
deformation technology to attain medium to high aspect ratio channels. The heat
transfer surface was made of aluminum.

The water was supplied to the system through a Neslab Chiller HX500 with an
operating temperature range of 5–35 °C. To protect the heat exchanger from
corrosion, 1 g potassium silicate solution per liter of water was added. The water
flow rate is measured through a water-side turbine flow meter, which is accurate to
1.5 %. An ABB K-5 flow meter was used for measuring the refrigerant-side flow
rate, and a 24 V DC pump was used for varying the refrigerant flow rate in the
system. A parallel plate condenser condenses the outlet refrigerant, which was
cooled by a Neslab M75 chiller. On the refrigerant side, the system was completely
evacuated before charging the refrigerant into the system. To ensure there is no
leakage, system pressure was monitored through the data acquisition system for an
entire day.

Fig. 3.25 Cooling capacity variation **a** with refrigerant flow rate for constant water mass flux; **b** with water flow rate for constant refrigerant mass flux

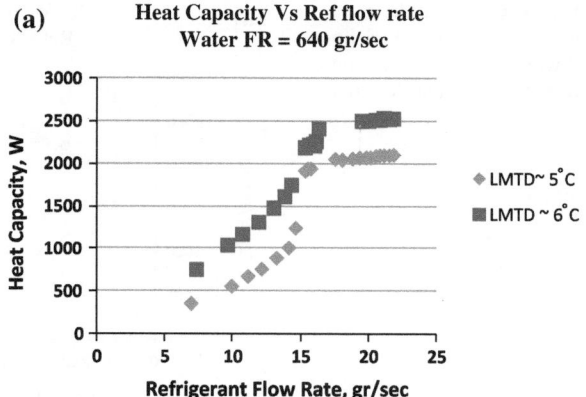

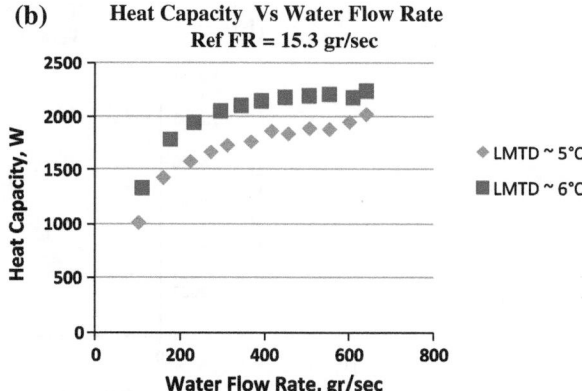

The evaporator test section had four T-type thermocouples at each measuring point—the inlet and exit for both the water and refrigerant sides. All thermocouples were calibrated within 0.05 °C. Differential pressure was also measured on both sides to measure the associated pressure drops. Absolute pressure measurement was required on the refrigerant side to determine the thermodynamic state of the refrigerant. All the measurements were recorded through the data acquisition system into a PC for constant monitoring and data logging.

Experimental data were collected for both constant water mass flow rate and constant refrigerant mass flow rate, as shown in Figs. 3.25a, b, respectively. Data are presented for two different sets of LMTDs (5 and 6 °C on average). Cooling capacity was observed to increase with increase in refrigerant flow rate in the lower mass flux range due to superheating of the vapor.

With variation in water mass flow rate, heat transfer rate was observed to increase for lower mass flux but to asymptote for medium and higher mass flux with a maximum uncertainty of ±15 %. Heat transfer in excess of 2 KW was observed at water mass flow rate from 200 to 640 ml/s for different LMTD values. Refrigerant mass flux was kept constant at 15.3 g/s.

Fig. 3.26 Overall heat
transfer coefficient variation
a with water mass flow rate;
b with refrigerant flow rate

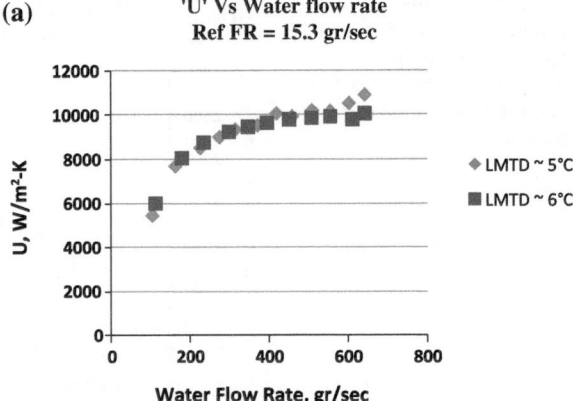

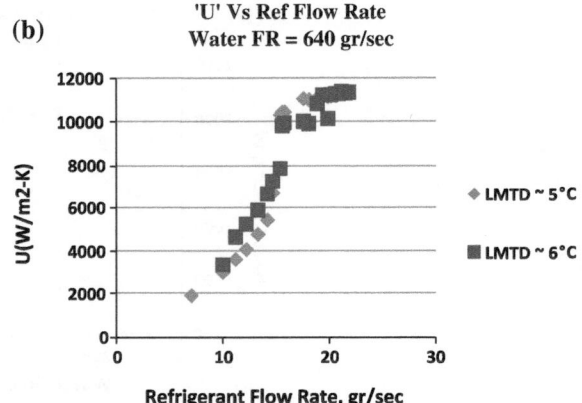

As shown in Fig. 3.26a, the overall heat transfer (U) coefficient increases with increase in water mass flow rate until it levels off at a mass flow rate of about 400 gm/s where it attains 10,000 W/m² K. With further increase in mass flow rate of water only modest improvement in the heat transfer coefficient was observed.

Variation of the overall heat transfer coefficient (U) with increasing refrigerant mass flow rate is shown in Fig. 3.26b. As seen there, U does seem to asymptote as water-side heat transfer coefficient levels off. Due to fabrication limitations, the inner-side channels (water side) were mini channels with hydraulic diameter of 600 μm. Pressure drop was found to increase with mass flow rate for both refrigerant and water sides. Pressure drop was in excess of 150 mbar on the water side at 640 ml/s and 60 mbar on the refrigerant side at 22 g/s, as shown in Fig. 3.27a and b, respectively. Water side pressure drop curve varies exponentially with increasing water side flow rate.

The heat exchanger was further modified and a high aspect ratio microgrooved tube was used with 110 μm hydraulic diameter for the grooves and with an enhancement insert on the water side. The modifications increased performance of

Fig. 3.27 Water side (**a**) and refrigerant side (**b**) pressure drop

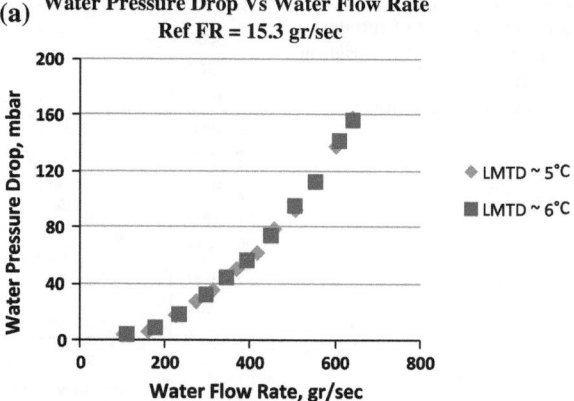

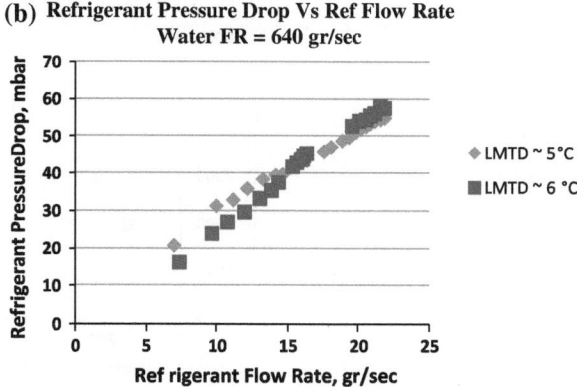

the heat exchanger while reducing its pressure drops. As shown in Figs. 3.28a, b, the heat transfer rate and the capacity of heat exchanger substantially increased, with the overall heat transfer coefficient reaching a level of 20,000 W/m^2 K (Fig. 3.29). Pressure drops on both sides of the heat exchanger were almost equal and on the order of 100 mbar (Fig. 3.30a, b) or less.

3.1.5.1   Manifolded Micro-Groove Condenser

Two-phase loops have many advantages over single-phase cooling loops in electronics cooling. Two-phase loops are more compact, require much less pumping power, are lighter and require much less working fluid. Two-phase loops also provide much greater heat transfer coefficients of evaporation and condensation resulting in the reduction of size and weight of heat exchangers, which is especially important in some applications, such as space/aerospace applications. The condenser described below utilizes the manifolded, micro-groove surface technology, which was described in Chap. 2 in detail. This design involves

Fig. 3.28 Variation of heat transfer capacity of tubular heat exchanger with variation of refrigerant (**a**) and (**b**) water flow

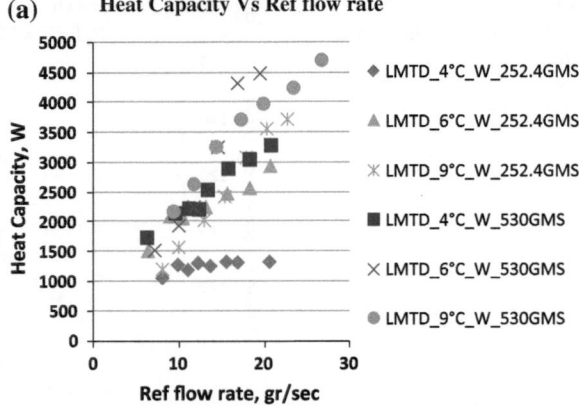

Fig. 3.29 Overall heat transfer coefficient of microgrooved surface tubular heat exchanger

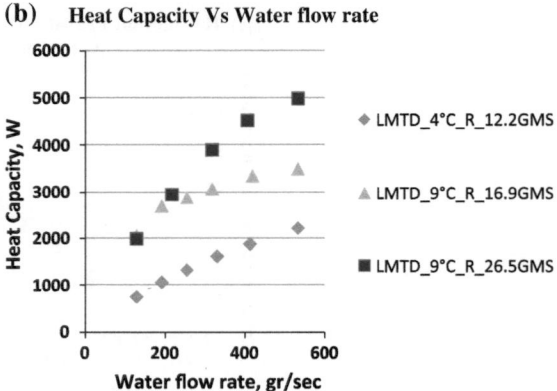

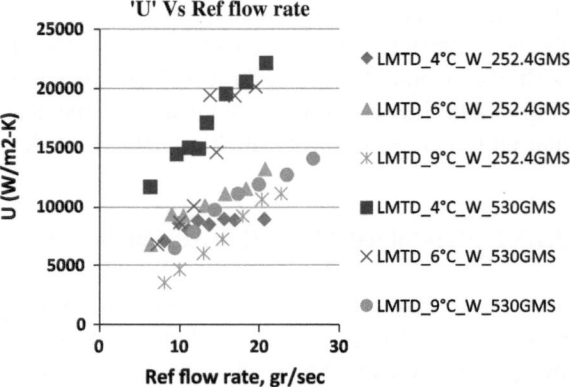

manifolds that force-feed the cooling fluid through narrow channels within the micro-structured surface for only a short distance. This novel concept combines the high heat transfer rates that are generated within narrow microchannels but

Fig. 3.30 Pressure drop of refrigerant side (**a**) and water side (**b**) of microgrooved surface tubular evaporator

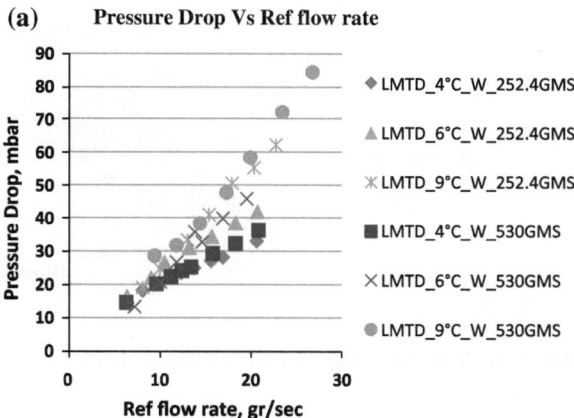

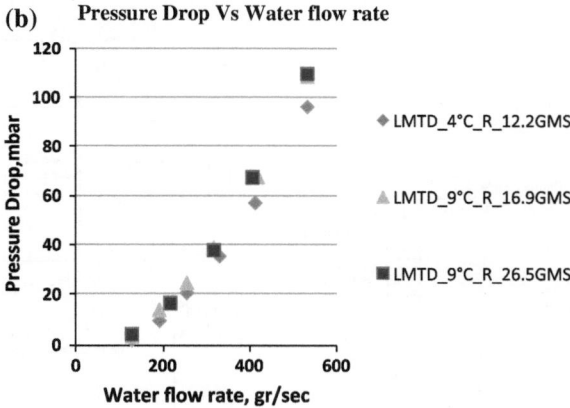

without the high pressure drops of conventional microchannel geometries. As described below this geometry also works well as a condenser for two-phase flow applications.

Condenser Design

A schematic diagram of a tubular micro-groove condenser is shown in Fig. 3.31. This condenser is a tubular variant of micro-grooved surface heat exchanger. A tubular configuration is compact, allows high pressure operation, and can be assembled without permanently bonding the heat exchanger components. The condensation process in the micro-grooves is identical in the cylindrical and rectangular geometries, since the curvature in the cylindrical tube is large compared with the depth of the micro-grooves. The tubular heat exchanger consists of a micro-grooved tube and internal and external flow distributors, which are placed in the cylindrical body of the heat exchanger and sealed with flanges.

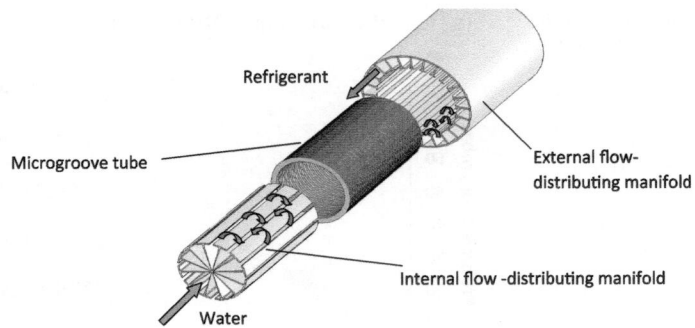

Fig. 3.31 Schematic diagram of a tubular condenser

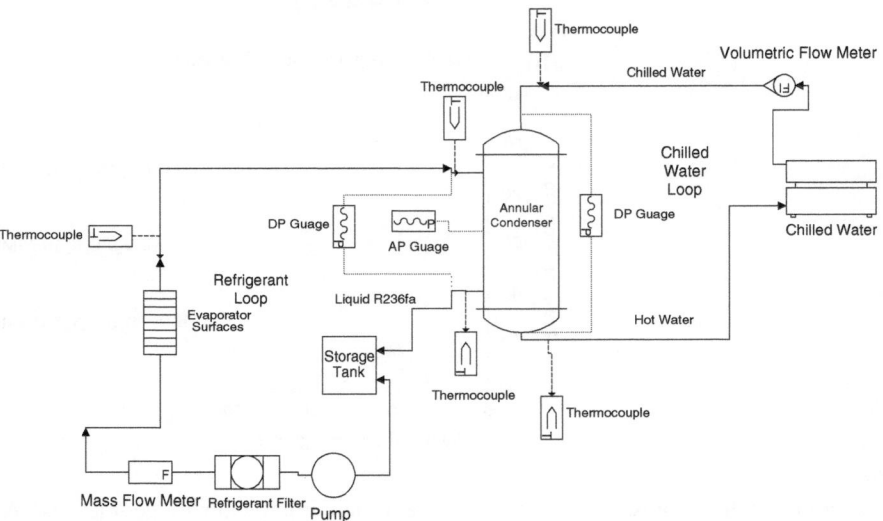

Fig. 3.32 Schematic diagram of R134a condenser test setup

Heat Transfer Testing Loop

The test loop, shown in Fig. 3.32, included the following components: evaporator, condenser, water chiller, storage tank, pump, mass flow meters (water and refrigerant), thermocouples, differential pressure sensor, and absolute pressure sensor. A photograph of the testing loop is shown in Fig. 3.33. The evaporator for the testing loop consisted of a thin-wall stainless steel tube heated directly by electrical current passing through it. The low electrical resistance of the metal requires electrical current in excess of 250 A; therefore, a high current transformer was used as a power source (see Fig. 3.34). The heater is in the bottom center part of the photograph with a storage tank attached to its housing.

Fig. 3.33 Photograph of the
new equipment to achieve
higher flow rates and
pressures

Fig. 3.34 Photograph of the
electrical heater and heating
coil

Determination of Refrigerant and Water-Side Heat Transfer Coefficients

In order to determine the heat transfer coefficient of the refrigerant side (h_{ref}), one
first needs to measure the overall thermal resistance, U,

$$\frac{1}{U*A} = \frac{1}{h_{ref}*A} + \frac{1}{h_{wtr}*A} \tag{3.1}$$

The total or overall heat transfer coefficient (U) is made up of the water side heat transfer coefficient (h_{wtr}), and refrigerant-side heat transfer coefficient h_{ref}. In order to calculate h_{ref} we must first calculate the other two. U is calculated using the following formula. (Fernandez-Seara et al. 2007):

$$U = Q_{wtr}/(A * LMTD)$$
(3.2)

where Q_{wtr} is the cooling load and is calculated by using the water-side data for a single-phase heat transfer process.

$$\dot{Q}_{wtr} = \dot{m}_{water} * c_{p_{water}} * (T_{waterout} - T_{waterin})$$
(3.3)

LMTD is the log mean temperature difference and is calculated as:

$$LMTD = \frac{T_{wout} - T_{win}}{\ln \frac{T_{win} - T_{sat}}{T_{wout} - T_{sat}}}$$
(3.4)

Here, h_{wtr} is only a function of $\dot{m}_{water}$, and since it is single phase and the temperature change is small, we can assume constant fluid properties. Thus, h_{wtr} can be assumed to be a function of Reynolds number or water velocity. A study was done to find the correlation between the h_{ref} and V_{wtr}. In order to do this h_{ref} needed to be kept constant. $h_{ref} = f(q'', T_{sat}, \dot{m}_{ref})$, so as long as the refrigerant mass flow rate, system pressure and Q_{out} were kept constant then h_{ref} would not vary. It was assumed that h_{wtr} is proportional to V^n, and Eq. (3.1) becomes (Fernandez-Seara et al. 2007):

$$\frac{1}{U} = K + C * \frac{1}{V^n}$$
(3.5)

The constant C has obtained the slope of the $1/U$ curve plotted against $1/V^n$. A range of U and V values were obtained from running a test matrix that varied water flow rate between 3 and 7 GPM and inlet water temperature between 14 and 18 °C.

The resulting data sets were grouped together at constant T_{sat}, $\dot{m}_{ref}$, and Q_{out} values, and each data set was graphed. The next step would be to determine the optimal value of 'n' in Eq. (3.5). If the value of n is correct then the standard deviation between the values of 'C' for each test would be ~ 0. By varying the values of n between 1.1 and 0.5 and plotting the standard deviation versus n we can see that there is an optimal point where the standard deviation is at a minimum as shown in Fig. 3.35.

Regression analysis would result in $n = 0.754$ with a value of $C = 0.0001185$ and the formula for $h_{wtr} = \frac{V^n_{wtr}}{C}$. Table 3.1 shows h_{wtr} versus mass flow rate of water.

The quality of the refrigerant entering the condenser also was estimated because h_{ref} is a function of the quality in the condenser. We based the inlet quality on the subcooled exit condition and the cooling capacity of the water side:

Fig. 3.35 Graph of standard
deviation of C versus n

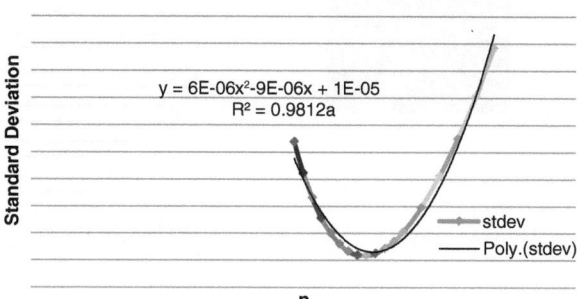

Table 3.1 Mass flow rate of water versus h_{wtr}

Volume flow rate (GPM)	Mass flow rate (kg/sec)	Velocity (m/s)	h_{wtr}
3	0.189	0.818	7249
4	0.252	1.090	9005
5	0.316	1.363	10656
6	0.379	1.635	12226
7	0.442	1.908	13733
8	0.505	2.180	15187
9	0.568	2.453	16598

$$X_{in} = (H_{in} - H_{out})/H_{fg} \qquad (3.6)$$

where H's are the enthalpies of the respective streams. H_{out} is known, H_{fg} is
calculated using the system pressure and qualities of $x = 1$ and 0, and H_{in} is
calculated using the following formula:

$$H_{in} = Q_{water}/m_{ref} + H_{out} \qquad (3.7)$$

Thermal Performance of a Tubular Micro-Groove Condenser

The flow manifold had the following dimensions: 2 mm width × 2 mm height
with 3 mm flow length. The details are shown in Fig. 3.36, and a photograph of the
manifold is shown in Fig. 3.37.

The condensation heat transfer coefficient as a function of saturation temperature is shown in Fig. 3.38. The refrigerant-side heat transfer coefficient increases
with saturation temperature because of the increase in vapor density with saturation temperature. The condenser pressure drop was only about 0.07 bar for all of
the tests (as shown in Fig. 3.39).

This condenser's high heat transfer coefficient would translate into a much
lighter, more compact condenser for a given application. The volume for a tubular

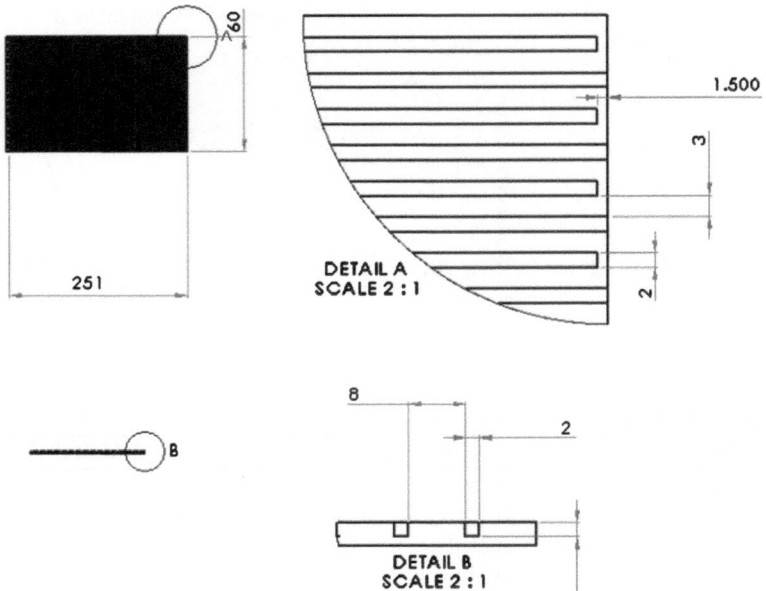

Fig. 3.36 Drawing manifold with 3 mm flow length

Fig. 3.37 Photograph of
manifold with 3 mm flow
length (edge view)

micro-groove condenser would be less than one-third (31 %) the volume of a state-
of-the-art conventional condenser. The weight of the tubular condenser would be
only about 38 % of a conventional condenser. Also, the pressure drop across the
cooling fluid-side of the condenser would be about one-third that of a conventional
fin-plate heat exchanger, which would in turn translate to reduce required operating power and energy losses.

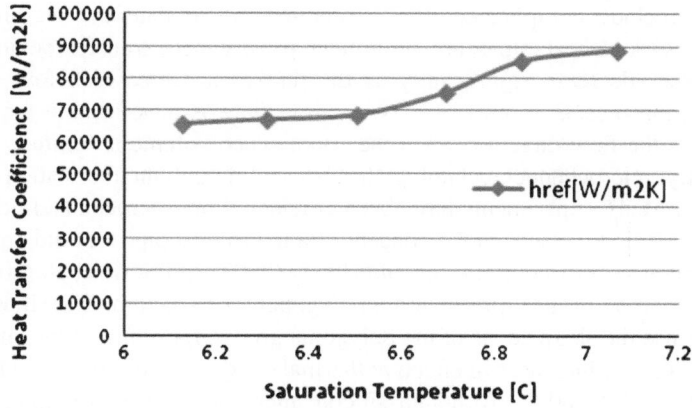

Fig. 3.38 Heat transfer coefficient as function of saturation temperature using new flow manifold with refrigerant R-134a

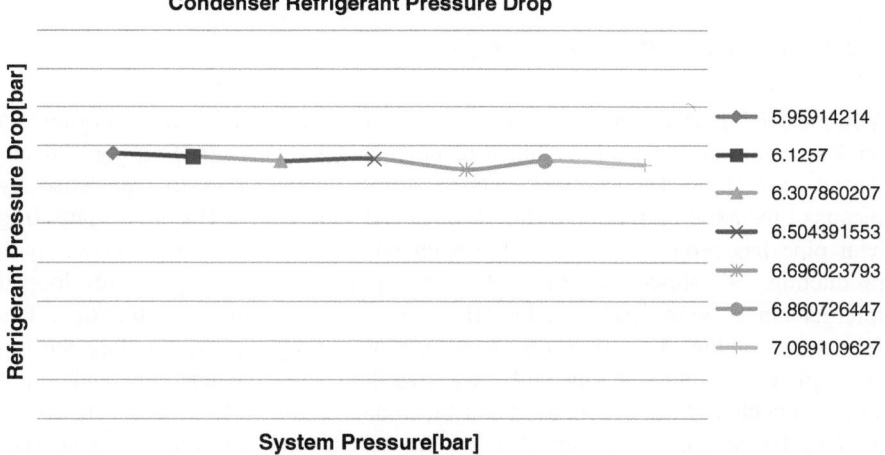

Fig. 3.39 Condenser refrigerant pressure drop as a function of system pressure with refrigerant R-134a

3.2 Microchannel Heat Pipes

Because advanced electronic equipment is decreasing in size, the circuit integration per unit area must increase, which in turn contributes to a rapid increase of heat generation rates. As a consequence, the operating temperatures of electronic components may exceed the desired temperature level, and if heat is not sufficiently removed, the failure rate of the equipment will cause an accelerated system failure. Due to the compactness of most modern electronic components, cooling devices also need to be small but highly effective in heat transport. Wicked heat pipes (or capillary-force-driven heat pipes) evaporate and condense the working fluid and include complicated wick structures to circulate the working fluid. Although wicked heat pipes are prominent passive heat transfer devices, their performances decrease significantly as the thickness decreases below a certain limit—for instance, 2–3 mm. This is because there is a noticeable temperature drop across the flow direction when the vapor space volume is limited.

Recently, microchannel technology has been applied to the fabrication of micro heat pipes. Many experiments have been conducted on microchannel heat pipes. Cotter (1984) first introduced the concept of a micro heat pipe that did not include a complicated wick structure. Kang and Huang (2002) and Berre et al. (2003) have studied silicon micro heat pipes with a polygonal cross-section. The silicon micro heat pipes use the sharp edges of the polygonal groove for return of the condensate to the evaporator. Increases in effective thermal conductivities compared to silicon ranged from 10 to 80 %. Although silicon micro heat pipes are small and can overcome the thickness limitations of wicked heat pipes, they cannot handle a large amount of heat, and their performance has yet to be improved.

3.2.1 Micro-Pulsating Heat Pipes

Youn and Kim (2012) developed a micro-pulsating heat pipe that did not have a wick structure and instead contained rectangular microchannels forming a meandering closed loop. The heat was transferred from an evaporator to a condenser by means of the axial oscillation of liquid slugs and vapor slugs. This micro-pulsating heat pipe has progressed beyond research laboratories and is now under mass production. As shown in Fig. 3.40, micro-pulsating heat pipes with looped microchannels were fabricated by MEMS or micro-machining technology. The fabrication of the micro-pulsating heat pipe was completed by bonding the top cover plate and filling the pipe with a working fluid. Then the heat input and output were connected at the evaporation and the condensation sections, respectively. A total of 10 parallel, interconnected rectangular channels forming a meandering closed loop were engraved on the silicon wafer with a thickness of 1 mm. The top of the silicon wafer was covered by a transparent glass plate with a thickness of 0.5 mm to allow visualization of the internal thermohydrodynamic behavior in the micro-pulsating heat pipes. The silicon wafer and glass plate were bonded together using anodic bonding. A hole of 1 mm diameter was drilled on the top of the glass plate to evacuate the micro-pulsating heat pipes and fill the micro-pulsating heat pipes with working fluid. Figure 3.40 represents the fabricated micro-pulsating heat pipes and dimensions of the heat pipe. The overall micro-pulsating heat pipes had a length of 50 mm, width of 15.5 mm, and thickness of 1.5 mm. The width and height of the engraved rectangular channel were 1 and 0.6 mm, respectively, and the hydraulic diameter was 0.75 mm. Ethanol was used as a working fluid. The micro-pulsating heat pipes achieved maximum effective thermal conductivity of 600 W/m K and a maximum heat transport capability of 4 W.

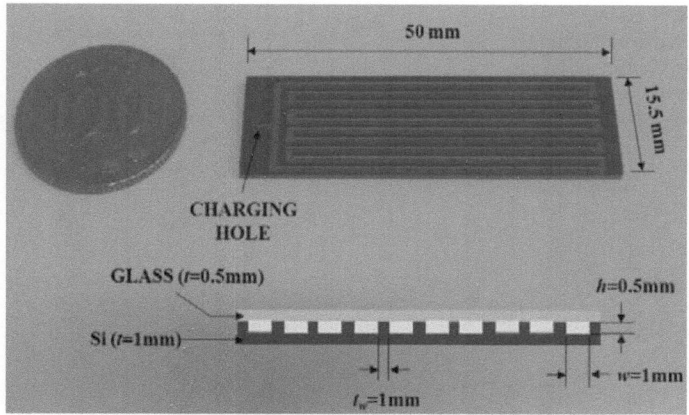

Fig. 3.40 Micro-pulsating heat pipe (Youn and Kim 2012)

As shown in Fig. 3.41a, the heat was transferred from the evaporator to the condenser sections by means of the axial oscillation of liquid slugs and vapor slugs. The micro-pulsating heat pipe is different from conventional heat pipes in design and working principle. There is no wick structure to return the condensed working fluid back to the evaporator section. The micro-pulsating heat pipe is made from a long meandering continuous capillary tube bent into many turns. The diameter of the tube is made sufficiently small that vapor plugs can be formed by capillary action as shown in Fig. 3.41b. The tube diameter that causes vapor plug and liquid slug arrangement (slug-train unit) is about 0.1–5 mm, and the critical diameter can be obtained (Akachi 1990) by

$$D_{crit} \leq 1.84 \sqrt{\frac{\sigma}{g(\rho_l - \rho_v)}} \qquad (3.8)$$

where σ is the surface tension, ρ_l is the liquid density, ρ_v is the vapor density, and g is the gravitational acceleration. If the tube diameter is too large, the liquid and vapor will tend to stratify. The micro-pulsating heat pipe operates in the following manner: the heat input increases the pressure of the vapor plug in the evaporator section; this pressure increase, in turn, pushes the neighboring liquid slugs toward the condenser section at low pressure. The heat is transported from the evaporator to the condenser by means of local axial oscillations and phase changes in the working fluids. A manufactured micro-pulsating heat pipe is shown in Fig. 3.42, in which the boundary between the liquid slugs and the vapor slugs is distinguishable.

The micro-pulsating heat pipe has many advantages, including high thermal performance and a maximum effective thermal conductivity of about 600 W/m·K, which is 3.5 times higher than that of silicon ($k = 162$ W/m·K) and 1.5 times that of copper ($k = 400$ W/m·K). It is possible to make a small, thin, and flat structure, and if a flexible material is used for the base and cover materials, it is possible to make a flexible micro-pulsating heat pipe because there is no wick structure.

Fig. 3.41 **a** Working
principle of micro-pulsating
heat pipe. **b** Vapor and liquid
arrangement according to
tube diameter (Youn and Kim
2012)

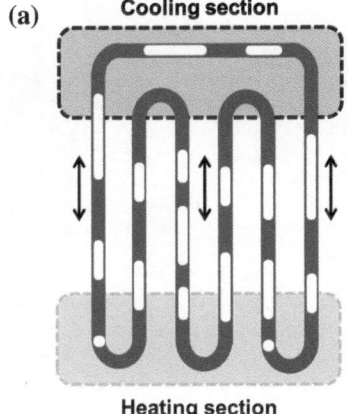

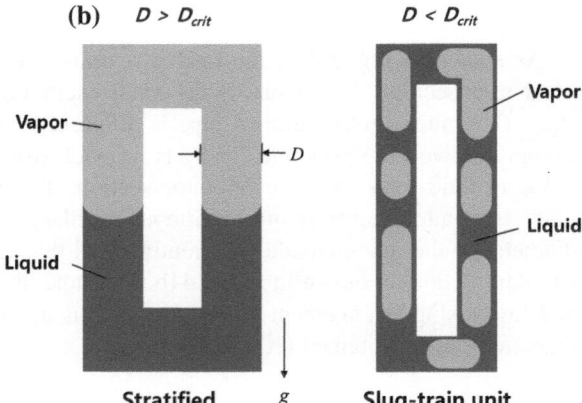

3.2.2 Microchannel Flat Heat Pipes

The potential devices for which the flat micro heat pipes fabricated may be applied
are mobile electronics such as small notebook PCs, PDAs, or cellular phones.
Although conventional heat pipes use either wire mesh or sintered metal for the
wick structure that is attached to the inner surface of a sealed metal housing, it is
difficult to adopt the same approach in micro flat heat pipes, as MHP size cannot be
reduced below a certain limit, for instance of 2–3 mm in thickness, due to the
volume of the wick structure itself. Thus, the flat micro flat heat pipes should
occupy as little space as possible to fit in the already small mobile device. Also,
since the main heat source in an electronic device is a chip, and the sides of most
chips are flat, the flat micro flat heat pipes are designed in a flat rectangular shape
with a thickness of 1.5 mm.

Recently, Lim et al. (2008) developed a micro heat pipe with a microchannel
groove wick structure using laser micro-machining technology. The structure of

Fig. 3.42 Top view of the micro-pulsating heat pipe with filling ratio of 50 % (Youn and Kim 2012)

Fig. 3.43 Structure of microchannel grooves (Lim et al. 2008)

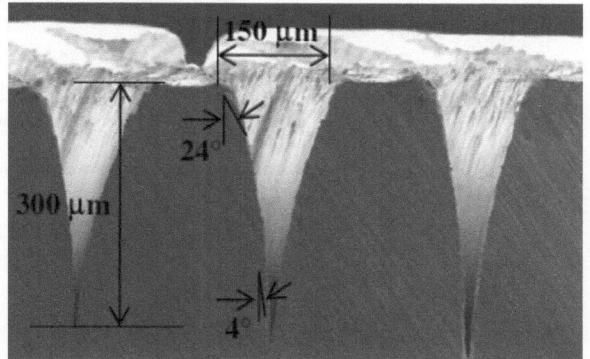

the microchannel grooves of a micro flat heat pipe is shown in Fig. 3.43. The microchannel grooves can be fabricated by laser micro-machining, MEMS, or LIGA technologies. The microgrooves of the micro flat heat pipe were fabricated on copper foil using a femtosecond laser micromachining technique. A commercial chirped pulse amplification laser system and a translation XYZ stage were used to fabricate the grooves under normal atmospheric air. The laser beam was focused on the workpiece using a 10x objective lens, and the workpiece was firmly attached on the translation stage using tape. The microgrooves, which serve to facilitate liquid passage during operation, were machined on the top and bottom plates. The width, depth, and length of each microgroove were about 150, 300, and 50 mm, respectively, and 16 grooves were produced on each plate.

As shown in Fig. 3.44, the fabrication of the micro flat heat pipe was completed by bonding the top cover plate and filling the pipe with a working fluid. Then the heat input and output were connected to the evaporation and the condensation sections, respectively. A manufactured micro-pulsating heat pipe is shown in Fig. 3.45. The working principle of the micro flat heat pipe is the same as that of a conventional heat pipe with a porous wick structure. The heat is transferred from an evaporator to a condenser. When one end of the heat pipe is heated, the working fluid inside the pipe at that end evaporates and increases the vapor pressure inside

Fig. 3.44 Structure of **a** the
overall micro flat heat pipe;
b the top and bottom plates
with microgrooves; and **c** the
middle plate for vapor
passage (unit: mm)

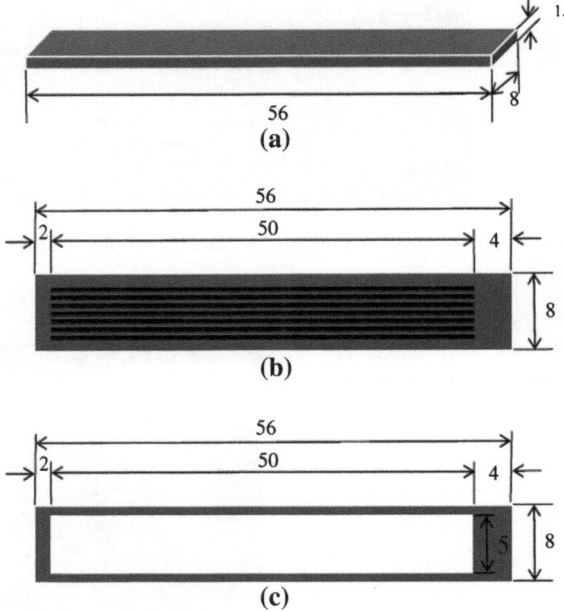

the cavity of the heat pipe. The latent heat of evaporation absorbed by the
vaporization of the working fluid reduces the temperature at the hot end of the heat
pipe. The heated vapor moves to the condensation section and condenses back into
a liquid at the cold interface, releasing the latent heat. The liquid then returns to the
hot interface through the capillary action of the microchannel grooves, where it
evaporates once more and repeats the cycle. The micro flat heat pipe has many
advantages, including its small, thin, and flat structure. It also has high thermal
performance, with a maximum heat transfer rate of 8 W under stable operation and
13 W at the dryout point, as shown in Fig. 3.46.

Recently, Ohadi et al. (2011) designed a microchannel surface-based heat
exchangers for a sodium heat-pipe solar energy receiver for a solar-powered
Stirling engine generator. Usually, the heat-pipe surface is made of very thin
inconel and is sintered very carefully to avoid defects. Any defects due to improper
sintering dry out that area of the heat pipe surface immediately due to concentrated
solar energy on the heat pipe receiver. In their current design, the micro-grooved
receiver surface was constructed of a solid nickel plate (as shown in Fig. 3.47),
and thus the micro-grooves cannot be separated from the heated surface. The
nickel plate also ensures faster, effective heat transfer without any dryout through
the sodium heat pipe from the receiver to the Stirling engine heater head.

Fig. 3.45   Top view of the micro flat heat pipe (Lim et al. 2008)

Fig. 3.46   Thermal resistance change at the dryout point of the micro flat heat pipe (Lim et al. 2008)

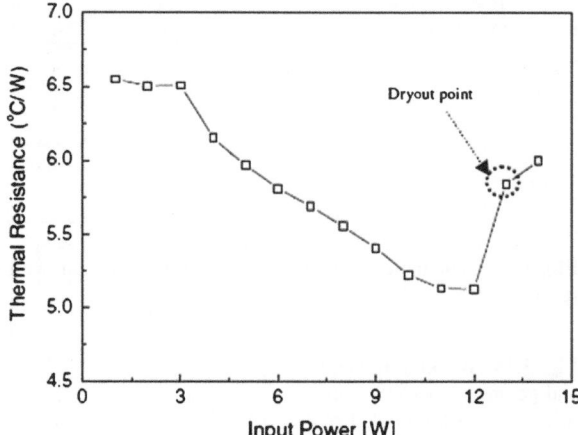

3.2.3   Counter-Stream-Mode Oscillating Flow Micro Heat Pipe

Nishio et al. (1995) proposed counter-stream-mode oscillating flow (COSMOS) heat pipes, wherein a meandering channel is employed to make an out-of-phase oscillating flow in neighboring channels. The COSMOS heat pipes can provide an enhanced heat diffusion effect compared with conventional heat pipes. Sugimoto et al. (2005) fabricated a micro-COSMOS heat spreader and verified that the theoretical model for the thermal performance of the COSMOS heat pipe proposed by Nishio et al. (1995) accurately predicted the thermal performance of the fabricated micro-COSMOS heat spreader. The heat spreader, fabricated by silicon micromachining, had 14 folded meandering Si grooves with total length of 700 mm, width of 575 μm, depth of 400 μm, and wall thickness between the adjacent channels of 150 μm. The heat spreader was connected to two

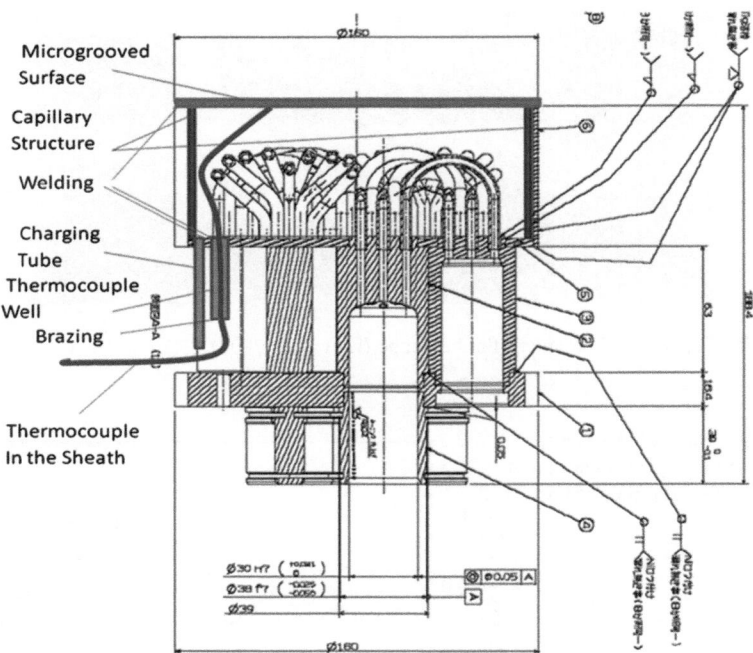

Fig. 3.47 Schematic of solar Stirling engine generator with micro-grooved sodium heat pipe receiver

Fig. 3.48 Working principle and geometry of COSMOS heat pipe (Nishio et al. 1995)

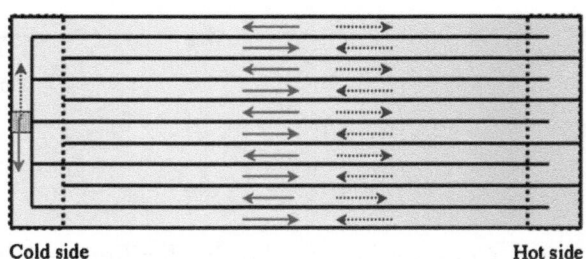

piezoelectric pumps. The total device size was 86×57 mm^2, and the effective cross-section of the heat pipe was 0.5×10 mm^2.

The operational principle and structure of the COSMOS heat pipe type heat spreader are shown in Fig. 3.48. Heat is transferred by thermal conduction between adjacent microchannels and does not rely on phase change. The fluid is driven by oscillation through densely arranged meandering microchannels placed between a heat source and a heat sink. The COSMOS operational principle is completely different from the loop heat pipe. Theoretically, the COSMOS heat pipe has higher effective thermal conductivity than the conventional heat pipes and

Fig. 3.49 Piezoelectric
pump of the COSMOS heat
pipe (Nishio et al. 1995)

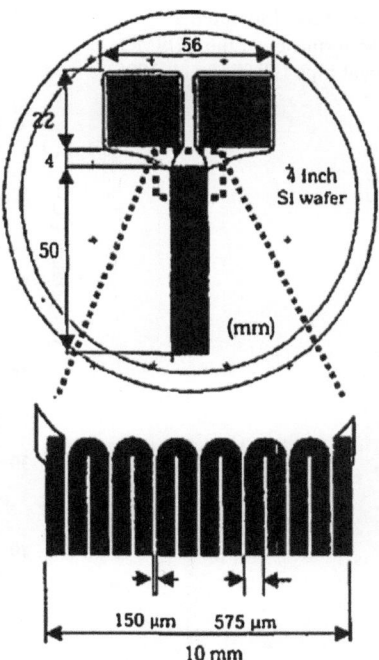

thus may be suitable for miniaturization of thermal equipment. Furthermore, the
cooling efficiency can be controlled continuously by controlling the oscillation
frequency or driving fluid volume. This is a unique feature of this device that
makes it possible to control the temperature of a heat source by varying the heat
spreading performance, transporting heat from the heat source to the heat sink.

The COSMOS heat pipe requires an additional external power source to sustain
the liquid oscillating motion. To realize a pump with the performance mentioned
above and a size thin enough to be eventually imbedded in a small space of an
electronic device, a dual piezoelectric pump configuration was adopted by Nishio
et al. (1995). Two pumps were connected to both ends of the channel, as sche-
matically shown in Fig. 3.49, and operated simultaneously to drive the fluid by
push–pull operation. The piezoelectric pump was constructed by bonding the PZT
plate with thickness of 120 μm and size of 20 mm on the Si diaphragm with
thickness of 100 μm and size of 22 mm. These thicknesses and sizes were based
on the FEM analysis. The COSMOS heat pipe demonstrated a temperature
gradient of 500 K/m and a heat transport rate of 20 W with water as the working
fluid. As shown in Fig. 3.50, the effective thermal conductivity was calculated as
9×10^3 W/mK, which is 23 times higher than that of copper.

Fig. 3.50 Measured
performance of the COSMOS
heat pipe (Nishio et al. 1995)

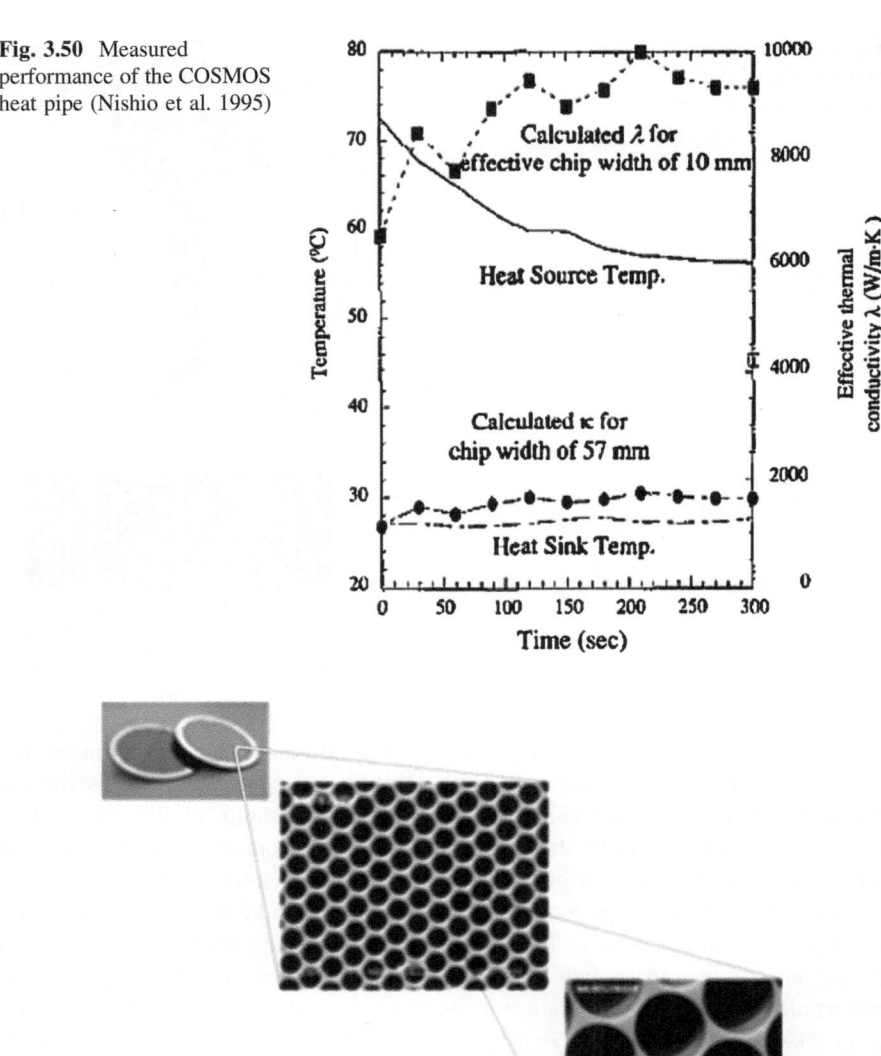

Fig. 3.51 Microchannel plate (Tectra Inc.)

3.3 Microchannel Heat Plates

A microchannel plate (MCP) is a planar component used to detect particles
(electrons or ions) and impinging radiation (ultraviolet radiation and X-rays), as
shown in Figs. 3.51 and 3.52. A typical MCP consists of about 10,000,000 closely
packed channels of common diameter, formed by drawing, etching, or firing in

hydrogen, forming a lead glass matrix. Typically, the diameter of each channel is ∼10 microns. Each channel acts as an independent, continuous dynode photomultiplier. MCPs are widely used to intensify low-light signal inside various image detectors. Industries for MCP applications range from astronomy to aerospace, machine-building, experimental physics, chemistry, biology, medicine, and ecology.

Authors' Biographies

Dr. Michael M. Ohadi is a professor of Mechanical Engineering and co-founder of the Center for Environmental Energy Engineering at the University of Maryland. He leads an industrial consortia in Advanced Heat Exchangers and Electronics Cooling Systems, with member companies from the U.S., Europe and Asia. He is internationally recognized for his research in enhanced heat and mass transfer, heat exchanger design and enhancement, advanced energy systems, and micro and nano applications of heat and mass transfer. He is the inventor/co-inventor of nine issued U.S. patents and has published close to 200 refereed technical publications. Dr. Ohadi has actively participated in promoting the higher education and support for fostering technological innovation and entrepreneurship. In 2002 he was featured in Washington Post for his efforts involving innovation and technology transfer. From 2003 to 2010 he was on leave assignment to the Petroleum Institute (Abu Dhabi) where served as Director of Mechanical Engineering program (2003–2006) and Provost and Acting President (2006–2010). He is a fellow member of both ASME and ASHRAE and has won numerous awards from both societies.

M. Ohadi et al., *Next Generation Microchannel Heat Exchangers*, 107
SpringerBriefs in Thermal Engineering and Applied Science,
DOI: 10.1007/978-1-4614-0779-9, © The Author(s) 2013

Dr. Kyosung Choo is a Research Associate in the Smart & Small Thermal Systems Laboratory in the mechanical engineering department at the University of Maryland, College Park. He received his Ph.D. from Korea Advanced Instituted of Science and Technology (KAIST) in 2011. His research interests include two-phase flow, jet impingement, energy efficiency in data centers, and electronics cooling. He has published numerous referred articles in his fields of expertise. He is listed in Who's Who in the World 2011 and a recipient of Outstanding Research Award from Brain Korea 21 in 2010 and Outstanding Academic Award from KAIST in 2011.

Edvin Cetegen received his Ph.D. degree in Mechanical Engineering from University of Maryland, College Park, in 2010. He is currently a member of the Fluids Group within Assembly Test and Technology Development Thermal & Fluids Core Competency, Intel Corporation, Chandler, AZ. His current research interests include thermal characterization and optimization of 3-D packaging, phase change heat transfer and thermal fluids analysis for the advanced package assembly process development.

Dr. Sergeui Dessiatoun is an Associate Research Professor of Mechanical Engineering at the University of Maryland, College Park. He has over 30 years of extensive experience in mechanical engineering design including design and development of thermal and hydraulic loops, aerospace and space cooling systems, environmental control systems, refrigeration and heat transfer systems, heat engines, diesel and gasoline fuel injection systems, and electronic control systems. Dr. Dessiatoun has been directing the ongoing research in the smart and Small Thermal Systems (S2TS) laboratory at the University of Maryland. He is the author/co-author of over 30 patents and more than 100 refereed articles in the area of energy transfer and conversion.

References

R. Abdallah, V. Meille, J. Shaw, D. Wenn, C. Bellefon, Gas-liquid and gas-liquid-solid catalysis in a mesh microreactor. Royal Society of Chemistry, **10**(4), 372–373 (2004)

H. Akachi, Structure of a heat pipe, U.S. Patent No. 4921041 (1990)

T.M. Adams, S.I. Abdel-Khalik, M. Jeter, Z.H. Qureshi, An experimental investigation of single-phase forced convection in microchannels. Int. J. Heat Mass Transf. **41**(6–7), 851–857 (1997)

B. Agostini, B. Watel, A. Bontemps, B. Thonon, Experimental study of single phase flow friction factor and heat transfer coefficient in minichannels, in *Proceedings of the Compact Heat Exchanger Symposium*, Grenoble, France, 2002

B. Agostini, J.R. Thome et al., High heat flux two-phase cooling in silicon multimicrochannels. IEEE Trans. Compon. Packag. Technol. **31**(3), 691–701 (2008)

Z. Anxionnaz, M. Cabassud, C. Gourdon, P. Tochon, Heat exchanger/reactors (HEX reactors): Concepts, technologies: State-of-the-art. Chem. Eng. Process. **47**, 2029–2050 (2008)

S. Baek, J. Kim, S. Jeong, Micro channel heat exchanger for LNG-FPSO application, in *Proceedings of 9th ISOPE Pacific/Asia offshore Mechanics Symposium*, Korea, 2010

Z.Y. Bao, D.F. Fletcher, B.S. Haynes, An experimental study of gas-liquid flow in a narrow conduit. Int. J. Heat Mass Transf. **43**(13), 2313–2324 (2000)

T. Baummer, E. Cetegen, M. Ohadi, S. Dessiatoun, Force fed evaporation and condensation utilizing advanced micro-structured surfaces and micro-channels. Microelectron. J. **39**, 975–980 (2008)

R. Beach, W.J. Benett, B.L. Freitas, D. Mundinger, B.J. Comaskey, R.W. Solarz, M.A. Emanuel, Modular microchannel cooled heat sinks for high power laser diode arrays. IEEE J. Quantum Electron. **28**(4), 966–976 (1992)

M.Le. Berre, S. Launay, V. Sartre, M. Lallemand, Fabrication and experimental investigation of silicon micro heat pipes for cooling electronics. J. Micromech. Microeng. **13**, 436–441 (2003)

A. Cavallini, D. Del Col, L. Doretti, M. Matkovic, L. Rossetto, C. Zilio, Two phase frictional pressure gradient of R236ea, R134a and R410A inside multi-port minichannels. Exp. Thermal-Fluid Sci. **29**, 861–870 (2005)

G.P. Celata, *Heat Transfer and Fluid Flow in Micro Channels* (Begell House, Inc., New York, 2004)

G.P. Celata, M. Cumo, A. Mariani, H. Nariai, F. Inasaka, Influence of channel diameter on subcooled flow boiling burnout at high heat fluxes. Int. J. Heat Mass Transf. **36**(13), 3407–3410 (1993)

G.O. Campbell, J.M. Fryer, S.M. Campbell, Enhanced Microchannel Cooler Development, Final Report to AFRL/PRTC, Contract F33615-03-C-2343, 30 Sept 2006

M. Ohadi et al., *Next Generation Microchannel Heat Exchangers*,
SpringerBriefs in Thermal Engineering and Applied Science,
DOI: 10.1007/978-1-4614-0779-9, © The Author(s) 2013

E. Cetegen, High heat flux cooling utilizing microgrooved surfaces. Mechanical Engineering. College Park, University of Maryland. PhD: 320, 2010

E. Cetegen, T. Baummer, S. Dessiatoun, M. Ohadi, *Heat Transfer Analysis of Microgrooved Evaporator and Condenser Surfaces Utilized in a High Heat Flux Two-Phase Flow Loop* (ASME International Mechanical Engineering Congress and Exposition, Seattle, 2007)

E. Cetegen, S. Dessiatoun, M. Ohadi, *Heat transfer analysis of force fed evaporation on microgrooved surfaces*. 6th International ASME Conference on Nanochannels, Microchannels and Minichannels. ICNMM2008-62285. Darmstadt, Germany, 2008

E. Cetegen, S. Dessiatoun, M. Ohadi, *Force Fed Boiling and Condensation for High Heat Flux Applications*. 7th Minsk International Seminar "Heat Pipes, Heat Pumps, Refrigerators, Power Sources", Minsk, Belarus, 2008

J.C. Chen, Correlation for boiling heat transfer to saturated fluids in convective flow. Ind. Eng. Chem. Res. **5**, 322–329 (1966)

D. Chisholm, Atheoretical basis for the Lockhart–Martinelli correlation for two-phase flow. Int. J. Heat Mass Transf. **10**, 1767–1778 (1967)

D. Chisholm, A.D.K. Laird, Two-phase flow in rough tubes. Trans. ASME **80**, 276–283 (1958)

K. Choo, S.J. Kim, Heat transfer and fluid flow characteristics of nonboiling two-phase flow in microchannels. ASME J. Heat Transf. **133**, 102901–102911 (2011)

K. Choo, S.J. Kim, Heat transfer and fluid flow characteristics of two-phase impinging jets. Int. J. Heat Mass Transf. **53**, 5692–5699 (2010a)

K.S. Choo, S.J. Kim, Comparison of thermal characteristics of confined and unconfined impinging jets. Int. J. Heat Mass Transf. **53**, 3366–3371 (2010b)

K.S. Choo, S.J. Kim, Heat transfer characteristics of impinging air jets under a fixed pumping power condition. Int. J. Heat Mass Transf. **53**, 320–326 (2010c)

K. Choo, T.Y. Kang, S.J. Kim, The effect of inclination on impinging jets at small nozzle-to-plate spacing. Int. J. Heat Mass Transf. **55**(2012), 3327–3334 (2012)

P.M.Y. Chung, M. Kawaji, The effect of channel diameter on adiabatic two-phase flow characteristics in microchannels. Int. J. Multiph. Flow **30**(7–8), 735–761 (2004)

P.M.-Y. Chung, M. Kawaji, A. Kawahara, Y. Shibata, Two-phase flow through square and circular microchannels—effects of channel geometry. J. Fluids Eng. **126**, 546–552 (2004)

J.G. Collier, Forced convection boiling. In *Two-Phase Flow and Heat Transfer in Power and Process Industries* (Hemisphere, Washington, 1981)

D. Copeland, Manifold microchannel heat sinks: analysis and optimization. Therm. Sci. Eng. **3**(1), 7–12 (1995a)

D. Copeland, Manifold microchannel heat sinks: numerical analysis. ASME Cooling Therm. Des. Electron. Syst. **15**, 111–116 (1995b)

D. Copeland, M. Behnia et al., Manifold microchannel heat sinks: Isothermal analysis. IEEE Trans. Compon. Packag. Manuf. Technol. Part A: **20**(2), 96–102 (1997)

T.P. Cotter, Principle and prospects for micro heat pipes, in *Proceedings of 5th International Heat Pipe Conference*, vol. 4 (1984), pp. 328–334

Y.L. Fan, L.G. Luo, Recent applications of advances in microchannel heat exchangers and multi-scale design optimization. Heat Transfer Eng. **29**, 461–474 (2008)

H.K. Forster, N. Zuber, Dynamics of vapor bubbles and boiling heat transfer. Chem. Eng. Prog. **1**(4), 531–535 (1955)

S.M. Ghiaasiaan, *Two-phase flow, boiling, and condensation in conventional and miniature systems* (Cambridge University, Cambridge, 2008)

C. Gillot, C. Schaffer, A. Bricard, Integrated micro heat sink for power multichip module. IEEE Trans. Ind. Appl. **36**(1), 217–221 (2000)

C. Gillot, C. Schaeffer, C. Massit, L. Meysenc, Double sided cooling for high power IGBT modules using flip chip technology. IEEE Trans. Compon. Packag. Technol. **24**(4) (2001)

V. Gnielinski, New equations for heat and mass transfer in turbulent pipe and channel flow. Int. Chem. Eng. **16**, 359–368 (1976)

N. Goel, D.Y. Goswami, Experimental verification of a new heat and mass transfer enhancement concept in a microchannel falling film absorber. J. Heat Transf. **129** (Feb 2007)

U. Grigull, H. Tratz, Thermischer einlauf in ausgebildeter laminarer rohrströmung. Int. J. Heat Mass Transf. **85**, 669–678 (1965)

K.E. Gungor, R.H.S. Winterton, A general correlation for flow boiling in tubes and annuli. Int. J. Heat Mass Transf. **29**, 351–358 (1986)

R. Hahn, A. Kamp, A. Ginolas, M. Schmidt, J. Wolf, V. Glaw, M. Topper, O. Ehrmann, H. Reichl, High power multichip modules employing the planar embedding technique and microchannel water heat sinks, in *Proceedings of IEEE 13th Semi-Therm Symposium*, 1997, pp. 49–56

D. Haller, P. Woias et al., Simulation and experimental investigation of pressure loss and heat transfer in microchannel networks containing bends and T junction. Int. J. Heat Mass Transf. (2009) (In review)

G.M. Harpole, J.E. Eninger, Micro-channel heat exchanger optimization, in *Proceedings of 7th IEEE Semi-Therm Symposium*, 1991

C. Harris, M. Despa, K. Kelly, Design and fabrication of a cross flow micro heat. J. Microelectromech. Syst. **9**(4), 502–508 (2000)

V. Hessel, P. Angeli, A. Gavriilidis, H. Lowe, Gas-liquid and gas-liquid-solid microstructured reactors: contacting principles and applications. Ind. Eng. Chem. Res. **44**, 9750–9769 (2005)

G. Hetsroni, A. Mosyak, E. Pogrebnyak, Z. Segal, Heat transfer of gas-liquid mixture in micro-channnel heat sink. Int. J. Heat Mass Transf. **52**, 3963–3971 (2009)

T.R. Hsu, A. Bar-Cohen, W. Nakayama, Manifold microchannel heat sinks: theory and experiment, in *Proceedings of the ASME International Electronic Packaging Conference*, vol. 2 (1995), pp. 829–835

Y.W. Hwang, M.S. Kim, The pressure drop in microtubes and the correlation development. Int. J. Heat Mass Transf. **49**, 1804–1812 (2006)

F.P. Incropera, D.P. DeWitt, *Fundamentals of Heat and Mass Transfer* (Wiley, New York, 2002)

K. Jahnisch, M. Baerns, V. Hessel, W. Ehrfeld, V. Haverkamp, H. Lowe, Ch. Wille, A. Guber, Direct fluorination of toluene using elemental fluorine in gas/liquid microreactors. J. Fluorine Chem. **105**, 117–128 (2000)

S.P. Jang, S.J. Kim, K.W. Paik, Experimental investigation of thermal characteristics for a microchannel heat sink subject to an impinging jet using a micro-thermal sensor array. Sens. Actuators, A **105**, 211–224 (2003)

N.R. Jankowski, L. Everhart, Comparing microchannel technologies to minimize the thermal stack and improve thermal performance in hybriz electric vehicles. IEEE 124–131 (2007)

V. Jha, M. Ohadi, S. Dessiatoun, A. Shooshtari, E. Al-Hajri, High performance micro-grooved evaporative heat transfer surface for low grade waste heat recovery applications, InterPACK 2011 Portland, Oregon (2011)

J. Jenks, V. Narayanan, Effect of channel geometry variations on the performance of a constrained microscale-film ammonia-water bubble absorber. J. Heat Transf. **130** (Nov 2008)

S.G. Kandlikar, A general correlation for two-phase flow boiling heat transfer coefficients inside horizontal and vertical tubes. ASME J. Heat Transf. **112**, 219–228 (1990)

S.G. Kandlikar, Fundamental issues related to flow boiling in minichannels and micro-channels. Exp. Therm. Fluid Sci. **26**, 389–407 (2002)

S.G. Kandlikar, High flux heat removal with microchannels—a roadmap of challenges and opportunities. Heat Transfer Eng. **26**(8), 5–14 (2005)

S.G. Kandlikar, A.V. Bapat, Evaluation of jet impingement, spray and microchannel chip cooling options for high heat flux removal. Heat Transfer Eng. **28**(11), 911–923 (2007)

S.G. Kandlikar, S. Garimella, D. Li, S. Colin, M.R. King, *Heat Transfer and Fluid Flow in Minichannels and Microchannels* (Elsevier, Amsterdam, 2006)

S.G. Kandlikar, W.J. Grande, Evolution of microchannel flow passages-thermohydraulic performance and fabrication technology. Heat Transfer Eng. **24**, 3–17 (2003)

S.W. Kang, D. Huang, Fabrication of star grooves and rhombus grooves micro heat pipe. J. Micromech. Microeng. **12**, 525–531 (2002)

A. Kawahara, P.M.Y. Chung, M. Kawaji, Investigation of two-phase flow pattern, void fraction and pressure drop in a microchannel. Int. J. Multiph. Flow **28**(9), 1411–1435 (2002)

W.M. Kays, M.E. Crawford, *Convective Heat and Mass Transfer* (McGraw-Hill, New York, 1980)

J.H. Kim, S. Baek, S. Jeong, J. Jung, Hydraulic performance of microchannel PCHE. Appl. Therm. Eng. **30**, 2157–2162 (2010)

Y.I. Kim, W.C. Chun, Forced air cooling by using manifold microchannel heat sinks. KSME Int. J. **12**(4), 709–718 (1998)

A. Kosar, Y. Peles, Boiling heat transfer in a hydrofoil-based micro pin fin heat sink. Int. J. Heat Mass Transf. **50**(5–6), 1018–1034 (2007)

R. Kukowski, *MDT- Micro deforamation Technology* (ASME IMECE, Washington D.C., 2003), 15–21 Nov. 2003

H.L. Langhaar, Steady flow in the transition length of a straight tube. J. Appl. Mech. **9**, A55–A58 (1942)

H.J. Lee, S.Y. Lee, Pressure drop correlations for two-phase flow within horizontal rectangular channels with small height. Int. J. Multiph. Flow **27**, 783–796 (2001)

D. Lelea, S. Nishio, K. Takano, the experimental research on microtube heat transfer and fluid flow of distilled water. Int. J. Heat Mass Transf. **47**(12–13), 2817–2830 (2004)

W. Li, Z. Wu, A general correlation for evaporative heat transfer in micro/minichannels. Int. J. Heat Mass Transf. **53**, 1778–1787 (2010)

H.T. Lim, S.H. Kim, H.D. Im, K.H. Oh, S.H. Jeong, Fabrication and evaluation of a copper flat micro heat pipe working under adverse-gravity orientation. J. Micromech. Microeng. **18**, 105013 (2008)

G. Maranzana, I. Perry, D. Maillet, Mini- and microchannels: influence of axial conduction in the walls. Int. J. Heat Mass Transf. **47**, 3993–4004 (2004)

E.D. Marquardt, R. Radebaugh, J. Dobak, A cryogenic catheter for treating heart arrhythmia. Adv. Cryog. Eng. **43**, 903–910 (1998)

P.M. Martin, W.D. Bennett, J.W. Johnston, Microchannel heat exchangers for advanced climate control. Proc. SPIE **2639**, 82–88 (1995)

G.D. Mathur, *Heat Transfer Coefficients and Pressure Gradients for Refrigerant R152a*, Alternate Refrigerant System Symposium, Pheonix, AZ, 2003

S.S. Mehendale, A.M. Jacobi, R.K. Ahah, Fluid flow and heat transfer at micro- and meso-scales with application to heat exchanger design. Appl. Mech. Rev. **53**, 175–193 (2000)

F.H. Mei, P.R. Parida, J. Jiang, W.J. Meng, S.V. Ekkad, Fabrication, assembly, and testing of Cu- and Al-based microchannel heat exchangers. J Microelectromech. Sys. **17**, 869–881 (2008)

K. Mishima, T. Hibiki, Some characteristics of air-water two-phase flow in small diameter vertical tubes. Int. J. Multiph. Flow **22**, 703–712 (1996)

L.J. Missaggia, J.N. Walpole, Z.L. Liau, R.J. Phillips, Microchannel heat sinks for two-dimensional high-power-density diode laser arrays. IEEE J. Quantum Electron. **25**(9), 1988–1992 (1989)

G.L. Morini, Single-phase convective heat transfer in microchannels: a review of experimental results. Int. J. Therm. Sci. **43**(7), 631–651 (2004)

I. Mudawar, Assessment of high-heat-flux thermal management schemes. IEEE Trans. Compon. Packag. Technol. **24**(2), 122–141 (2001)

D. Mundinger, R. Beach, W. Benett, R. Solarz, W. Krupke, R. Staver, D. Tuckerman, Demonstration of high performance silicon microchannel heat exchangers for laser diode array cooling. Appl. Phys. Lett. **53**(12), 1030–1032 (1988)

D.M. Murphy, B. Rosen, J. Blasi, N.P. Sullivan, R.J. Kee, M. Hartmann, N.E. McGuire, Ceramic microchannel heat exchanger and reactor for SOFC applications. ECS Trans. **35**(1), 2835–2843 (2011)

G.F. Nellis, A heat exchanger model that includes axial conduction, parasitic heat loads, and property variations. Cryogenics **43**, 523–538 (2003)

E.Y.K. Ng, S.T. Poh, Investigative study of manifold microchannel heat sinks for electronic cooling design. J. Electron. Manuf. **9**(2) (1999)

N.-T. Nguyen, S.T. Werely, *Fundamentals and Applications of Microfluidics* (Artech House, Boston, 2002)

S. Nishio, X.H. Shi, W.M. Zhang, Oscillation-induced heat transport: heat transport characteristics along liquid-columns of oscillation-controlled heat transport tubes. Int. J. Heat Mass Transf. **38**, 2457–2470 (1995)

M. Ohadi, S. Dessiatoun, A. Shooshtari, J. Kelley, J. Fody, A project report on design and analysis of high temperature sodium heat pipe installed on a concentrated solar receiver stirling engine head (2011)

J.H. Ryu, D.H. Choi, Three dimensional numerical optimization of a manifold microchannel heat sink. Int. J. Heat Mass Transf. **46**, 1553–1562 (2003)

B. Palm, *Heat Transfer in Microchannel*, Proceedings of Heat Transfer and Transport Phenomena in Microchannel (Begell House Inc., Banff, Canada, 2000), pp. 54–64, ISBN 1-56700-150-5

M. Parrino, A. Parola, L. Dentis, A high efficiency mechanically assembled aluminum radiator with real flat tubes, SAE Tech. Paper Series 940 495 (1994)

X.F. Peng, G.P. Peterson, Convective heat transfer and flow friction for water flow in microchannels structures. Int. J. Heat Mass Transf. **36**(12), 2599–2608 (1996)

X.F. Peng, B.X. Wang, Forced convection and fluid flow boiling heat transfer for liquid flowing through microchannels. Int. J. Heat Mass Transf. **36**(14), 3421–3426 (1993)

X.F. Peng, B.X. Wang, G.P. Peterson, H.B. Ma, Experimental investigation of heat transfer in flat plates with rectangular microchannels. Int. J. Heat Mass Transf. **38**(1), 127–137 (1995)

J. Pfahler, J. Harley, H. Bau, J. Zemel, Liquid transport in micron and sub micron channels. Sens. Actuators **A21–23**, 431–434 (1990)

J. Pfahler, J. Harley, H. Bau, J. Zemel, Gas and liquid flow in small channels. Micromech. Sens. Actuators Syst. ASME **DSC-32**, 49–60 (1991)

R.J. Philips, Microchannel heat sink. Lincoln Lab. J. **1**(1), 31–48 (1988)

V. Ponyavin, Y.T. Chen, A.E. Hechanova, M. Wilson, Numerical modeling of compact high temperature heat exchanger and chemical decomposer for hydrogen production. Heat Mass Transf. **44**, 1379–1389 (2008)

B. Schilder, S.Y.C. Man, N. Kasagi, S. Hardt, P. Stephan, Flow visualization and local measurement of forced convection heat transfer in a microtube. ASME J. Heat Transf. **132**, 031702 (2010)

R. Schmidt, Challenges in electronic cooling: opportunities for enhanced thermal management techniques-microprocessor liquid cooled minichannel heat sink, in *First International Conference on Microchannels and Minichannels*, Rochester, NY, 24–25 Apr 2003, pp. 951–959

C. Schmitt, D.W. Agar, F. Platte, S. Buijssen, B. Pawlowski, M. Duisberg, Ceramic plate heat exchanger for heterogeneous gas-phase reactions. Chem. Eng. Technol. **28**, 337–343 (2005)

A. Serizawa, Z. Feng, Z. Kawara, Two-phase flow in micro-channels. Exp. Therm. Fluid Sci. **26**, 703–714 (2002)

M.M. Shah, A new correlation for saturated boiling heat transfer: Equations and further study. ASHRAE Trans. **88**(1), 185–196 (1982)

M.M. Shah, Evaluation of general correlations for heat transfer during boiling of saturated liquids in tubes and annuli, J. HVAC&R Res. **12**(4), 1047–1064 (2006)

R.K. Shah, A.L. London, *Laminar Flow Forced Convection in Ducts* (Academic, New York, 1978)

E.N. Sieder, G.E. Tate, Ind. Eng. Chem. **28**, 1429 (1936)

A. Sommers, Q. Wang, X. Han, C. T'Joen, Y. Park, A. Jacobi, Ceramics and ceramic matric composites for heat exchangers in advanced thermal systems—a review. Appl. Therm. Eng. **30**, 1277–1291 (2010)

M. Sugimoto, K. Minai, M. Uemura, S. Nishio, O. Tabata, A novel micro counter stream mode oscillating flow heat pipe, in *Proceedings of 18th IEEE International Conference on MEMS*, vol. 2 (2005), pp. 606–609

M.K. Sung, I. Mudawar, CHF determination for high-heat flux phase change cooling system incorporating both micro-channel flow and jet impingement. Int. J. Heat Mass Transf. **52**(3–4), 610–619 (2009)

M. Suo, P. Griffith, Two-phase flow in capillary tubes. J. Basic Eng. **86**, 576–582 (1964)

P. Stephan, Microscale evaporative heat transfer: modelling and experimental validation, in *12th Heat Transfer Conference*, Grenoble, France, 2002

P. Thors, N. Zoubkov, Tool for Making Enhanced Heat Transfer Surfaces, USA, 2009

D.B. Tuckerman, R.F. Pease, High performance heat sinking for VLSI. IEEE Electron. Dev. Letts. **EDL-2**, 126–129 (1981)

K.A. Triplett, S.M. Ghiaasiaan, S.I. Abdel-Khalik, D.L. Sadowski, Gas-liquid two phase flow in microchannels, Part 1: two-phase flow patterns. Int. J. Multiph. Flow **25**, 377–394 (1999)

US Energy Report, Waste Heat Recovery: Technology and Opportunities in US Industries (2008)

A.V. Viday, S.B. Soshelev, G.V. Reznikov, V.V. Kharitonov, S.V. Cheremushkin, Thermophysical design of the parameters of a computer board with a microchannel cooling system. J. Eng. Phys. Thermophys. **64**(1), 80–86 (1993)

M. Visaria, I. Mudawar, Theoretical and experimental study of the effects of spray inclination on two-phases spray cooling and critical heat flux. Int. J. Heat Mass Transf. **51**(9–10), 2398–2410 (2008)

Waldmann, Evaluation of process systems for floating LNG production units, Tekna Conference 18, 2008

B.X. Wang, X.F. Peng, Experimental investigation on liquid forced convection heat transfer through microchannels. Int. J. Heat Mass Transf. **37**(1), 73–82 (1994)

R. Webb, P. Farrell, Improved thermal and mechanical design of copper/brass radiators, SAE Tech. Paper Series 900 724 (1990)

P.Y. Wu, W.A. Little, Measurement of friction factor for the flow of gases in very fine channels used for microminiature Joule Thompson refrigerators. Cryogenics **23**(5), 273–277 (1983)

P.Y. Wu, W.A. Little, Measurement of heat transfer characteristics of gas flow in fine channels heat exchangers used for miniature refrigerators. Cryogenics **24**(5), 415–420 (1984)

G. Xia, Q. Liu, Influence of surfactant on friction pressure drop in manifold microchannels. Int. J. Therm. Sci. **47**, 1658–1664 (2008)

B. Xu, K.T. Ooi, N.T. Wong, W.K. Choi, Experimental investigation of flow friction for liquid flow in microchannels. Int. Comm. Heat Mass Transf. **27**(8), 1165–1176 (2000)

L.P. Yarin, A. Mosyak, G. Hetsroni, *Fluid flow, heat transfer and boiling in micro-channels* (Springer, Berlin, 2009)

T.-H. Yen, N. Kasagi, Y. Suzuki, Forced convective boiling heat transfer in microtues at low mass and heat fluxes. Int. J. Multiph. Flow **29**, 1771–1792 (2003)

Y.J. Youn, S.J. Kim, Fabrication and evaluation of a silicon-based micro pulsating heat spreader. Sens. Actuators, A **174**(2012), 189–197 (2012)

D. Yu, R. Warrington, R. Barron, T. Ameel, An experimental investigation of fluid flow and heat transfer in microtubes, in *Proceedings of the ASME/JSME Thermal Engineering Conference, ASME*, vol. 1 (1995) pp. 523–530

J. Yue, L. Luo, Y. Gonthier, G. Chen, Q. Yuan, An experimental investigation of gas-liquid two-phase flow in single microchannel contactors. Chem. Eng. Sci. **63**, 4189–4202 (2008)

L. Zang, K.E. Goodson, T.W. Kenny, *Silicon Microchannel Heat Sinks* (Springer, Berlin, 2003)

H. Zhang, G. Chen, J. Yue, Q. Yuan, Hydrodynamics and mass transfer of gas-liquid flow in a falling film microreactor. AIChe J. **55**(5) (2009)

T.S. Zhao, Q.C. Bi, Co-current air-water two-phase flow patterns in vertical triangular microchannels. Int. J. Multiph. Flow **27**, 765–782 (2001)